Reality, Science and the Supernatural

CAN SCIENCE SUPPORT BELIEF IN A CREATOR GOD?

John Brain

To my wife Joyce for her invaluable support.

CONTENTS

ACKNOWLEDGMENTS i

PREFACE 1

CHAPTER 1. The Project 10

   1.1 Introduction 11
   1.2 Main Doubts 14
     Materialism and the mind/matter relationship *14*
     Evolution *18*
     Scientific and Personal Explanation *21*
     Supporting evidence from science *24*
   1.3 A Project Outline 26
   1.4 Key Points 29

CHAPTER 2. Materialism and Reality 30

   2.1 Introduction 31
   2.2 Modern Physical Sciences 34
   2.3 The Nature of Matter 36
     2.3.1 Looking inwards *36*
     2.3.2 Looking upwards *52*
   2.4 Discussion 54
   2.5 Key Points 57

CHAPTER 3. Evolution and Reality 59

   3.1 Introduction 60
   3.2 Darwinian Evolution 62
   3.3 Three Viewpoints on Human Development 64
     Young Earth Creationism *64*
     Intelligent Design *66*
     Evolution *68*

3.4 Does belief in God prevent acceptance of the theory of evolution?..................................... 72

3.5 Does evolution need God?................................. 76

3.6 Theistic Evolution 79

What is theistic evolution?.................................. 79

Why can evolution be such a cruel process?........ 81

Does Theistic Evolution introduce a "God of the Gaps"? 82

3.7 Discussion 83

3.8 Key Points 84

CHAPTER 4. Science and Reality 86

4.1 Introduction 87

4. 2. Advances in science – emergence of a new Reality 89

Biology.. 89

Physical Sciences 96

4.3 A More Reliable Explanation of Reality 99

Some dogmas of modern science........................ 101

No 'theory of everything'.................................... 103

No satisfactory explanation of quantum physics 105

More intriguing problems.................................... 106

4.4 Limitations of Science...................................... 108

Science is continually changing 109

Science is not geared to answer 'why' questions 110

The Laws of science have no creative power 110

Science does not make moral judgments............ 112

Science does not tell us how to use knowledge... 112

4.5 Discussion 113

4.6 Key Points 115

CHAPTER 5. Signposts from Science to Reality....... 118

5.1 Introduction 119
5.2 The Order of the Universe – God's Laws at work?
.... 120
5.3 The Goldilocks Effect...................................... 122
The value of entropy at the start of the universe 123
Production of carbon in the stars....................... 126
Dark energy and dark matter 127
Chance or Design?... 129
5.4 The Human Genome Project 134
The Language of God.. 134
5.5 The Intelligibility of the Universe 139
5.6 Discussion.. 142
5.7 Key Points 145

CHAPTER 6. Adequate Evidence............................ 147

6.1 Introduction 148
6.2 Materialism – a flawed and outdated philosophy 150
Small scale world – quantum mechanics............ 151
Self-organising systems – Chaos Theory............ 153
Discussion... 154
6.3 Theistic evolution provides the best available
explanation 155
6.4 Is science fully equipped to answer "The God
Question"? 156
Our scientific knowledge is continually changing
.. 157
Scientists have yet to provide a satisfactory
explanation for the quantum mechanics............. 157
Science on its own cannot answer the "God
Question" 158

6.5 Evidence to support belief in a creator God 159

The order of the Universe *159*

The Human Genome *160*

The Intelligibility of the Universe *161*

Dark Energy *161*

The Goldilocks effect *162*

6.6 Looking to the future – another new reality? ... 163

The Holographic Universe *165*

6.7 Science and Philosophy 168

6.8 Conclusions 170

APPENDIX 1. Suffering and Evolution 173

APPENDIX 2. The Moral Law – The Start of My
Christian Journey 178

REFERENCES 191

INDEX 204

ACKNOWLEDGMENTS

This book was written as a contribution to the Scientists in Congregations Scotland Programme centred at the University of St Andrews and led by Dr Andrew Torrance. I am very grateful to Andrew for the excellent support and assistance he has given me in completing the work. His proof-reading of several chapters was meticulous and his comments, advice and encouragement proved immensely supportive.

I would also like to thank my friend, Dr Laura Bence from Cadzow Parish Church, Hamilton, for her great support in our SICS project.

PREFACE

Scientists investigate that which already is; Engineers create that which has never been.
- ***Albert Einstein*** *(1879-1955)*

(Faith) affects the whole of man's nature. It commences with the conviction of the mind based on adequate evidence; it continues in the confidence of the heart or emotions based on conviction, and it is crowned in the consent of the will, by means of which the conviction and confidence are expressed in conduct."
- ***W H Griffith Thomas*** *(Anglican Theologian 1861-1924)*

For most of my adult life I was involved in undertaking, leading and eventually overall management of research and development programmes in the United Kingdom's National Engineering laboratory. I was very involved in my work and it was not until fairly recently, when I fully retired, that I took time to stop and make an assessment of my faith.

At present the popularity of the Christian church

here in Scotland is falling rapidly. While selfish materialism and apathy to the teachings of the church are prominent among the population and contribute greatly to this decline, it is also true, in my experience, that an increasing number of people in the United Kingdom are being led by reasoned argument to consider that atheism provides the answer. This seems to be particularly popular among scientists. Indeed it was the writings of the eminent scientist and evolutionist, Richard Dawkins, particularly in his book *The God Delusion (1)* that eventually spurred me to initiate a reassessment of my views on the relationship between science and religion.

Professor Dawkins is an exceptionally motivated and talented scientist. He argues his case for Darwinism with an infectious enthusiasm and, when he sticks to science, reading his work is an absolute pleasure. He leaves you with little doubt of the strength of his arguments and I can see why he is so highly committed to his scientific effort. At present I am reading a copy of his book *The Greatest Show on Earth (2)* and it is proving to be as enjoyable and informative as I had hoped. I am more convinced than ever by the arguments for evolution.

However, when he turns his attention to religion Professor Dawkins' arguments prove much less convincing and this is not helped when he decides to adopt what Alister McGrath, Professor of Theology at King's College London, refers to as "turbocharged rhetoric and highly selective manipulation of facts", *(4)*. Nevertheless his efforts did cause me to question the strength of my faith and as a result of that questioning decide that a reassessment of my position was badly needed. In addition, for a long time now I have felt that the "packaging" of the essential messages of Christianity is grossly out of date. I believe that, while the fundamental teachings must remain unchanged, development of the faith is urgently needed. Those who insist on a literal interpretation of the Bible do their cause immeasurable harm and their output provides excellent cannon fodder for the outrageous onslaughts of Professor Dawkins.

Excellent responses to Richard Dawkins' works have been given by Alister McGrath in *Dawkins' God (3)* and more recently, along with his wife Joanna, in *The Dawkins Delusion (4)*, John Lennox in, *God's Undertaker. Has Science Buried God? (5)*, and Keith Ward, *Why There Almost Certainly is a God (6)*. For the scientist or engineer interested in religion I would also warmly recommend *The Language of God – A Scientist Presents Evidence for Belief (7)* by Francis Collins. Dr Collins is one of the world's leading scientists and was the director of the highly successful Human Genome Project *(8)*. However, I would like to make clear here that my main aim in writing this work was not simply to air my disagreements with Professor Dawkins and his atheist friends but to update my views on science and religion taking into account major recent advances in knowledge, particularly in the physical sciences. I hoped that this would help to affirm, at least in my mind, my belief in the supernatural and perhaps help others wrestling with the same problems.

While I have read a number of books on "Science and Religion" I have not yet come across one where a research engineer has provided the scientific input. So another reason for writing this book was to put this right. I believe that we research engineers can bring a fresh approach to the discussion and can be readily sympathetic to certain views expressed in both camps.

In the United Kingdom the popular view of the modern engineer, like the popular view of the modern Christian, is very outdated. Currently research engineers perform leading edge investigations in a large number of areas, for instance we design rockets and send sophisticated probes out to explore the universe. These space probes are, like the Hubble telescope, masterpieces of engineering design and ingenuity. Software engineers are at the heart of computing advancement and the information technology explosion which is revolutionising our lives. Even in the field of *quantum computing* engineers lead the research work and it is pleasing to note that the distinguished leading researcher in quantum physics Seth Lloyd *(9)*, a professor of mechanical engineering at the Massachusetts Institute of Technology in the USA, likes to refer to himself as a "quantum mechanic". Then, in addition to the well-recognised engineering research going on in the aircraft, shipping and motor car industries there are also biochemical engineers and of course genetic engineers. I could go on but it is manifestly clear that the popular notion of a typical engineer being like Dan McPhail the ship's engineer in *Para Handy (10)* who is "having trouble with his boilers", is, perhaps, not quite right. Hence we engineers can have considerable sympathy with modern Christians where the common view of their beliefs, such as that expressed by Richard Dawkins, is equally outdated.

In several ways engineers are different from pure scientists and I believe that we can help to freshen up the discussion on "Science and Religion". To support this statement I would like to illustrate a few of these differences using three examples. Firstly, while we are basically scientists, engineers are more than often concerned with practical problems and have to give priority to finding practical solutions. So, unlike many of our scientific colleagues, we are happy to press ahead using empirical formulae without waiting for exacting theoretical verifications. To find solutions we often use methods which incorporate techniques such as *dimensional analysis* to help reduce highly complex problems to manageable proportions. Some idea of the difference in the approach of an engineer and say, a pure mathematician, can be illustrated by the following example:-

A beautiful lady was standing at the edge of a field watching the arrival of two men, an engineer and a mathematician, at the opposite edge. On seeing the men the lady shouted to them that if either of the gentlemen would like to cross the field she would give a kiss to the one who chose to come to her. However, she stipulated that he had to walk in a straight line and he could only step towards her in increments where the length of each increment was half of the distance between her and where he started the increment. Immediately the mathematician thought, *Simple maths shows me that there is no way I could ever reach this lady using the stipulated conditions.* However, while he was thinking the engineer had started moving as instructed and when he had come within an appropriate distance of the lady he decided, *That's near enough,* and she leaned forward and gave him his rewarding kiss.

While I feel sure that the mathematician would very quickly realise his mistake, I feel equally sure that his immediate reaction would not be the immediate practical reaction of the engineer!

Secondly, innovation is a big priority for members of the engineering profession. The vast majority of the inventions that have marked the tremendous advances in technology over the last century have been introduced by engineers, and if I decided to describe them I am sure that I could fill many substantial volumes. Albert Einstein emphasised the clear difference between scientists and engineers by stating that: "Scientists investigate what already is; Engineers create what has never been" *(11)*. I have been particularly impressed by the number of collaborative inventions which have involved the blending together of numerous disciplines to achieve the main objective. An excellent example of this is the satellite navigation system now used in most of our motor cars. This little box of tricks sits in your car and bounces pulses of ultrasound off a satellite orbiting the earth. These signals are then used with amazing accuracy to pinpoint your position on the planet and lead you to your destination with unerring accuracy, pointing out any difficulties on the way (if you have installed the latest software!). I would invite you to consider this device and list the number of disciplines involved in its successful operation. I believe that you will be surprised by the largeness of the number. Interdisciplinary innovation is a major strength for engineers and we often have to work with data not related to the physical sciences. N. W. Dougherty, a professor of Civil Engineering at the University of Tennessee, contends: "The ideal engineer is a composite...He is not a scientist, he is not a mathematician, he is not a sociologist or writer... but he may use the knowledge and techniques of any or all of these disciplines in solving engineering problems" *(12)*.

Finally, if we consider an artist as "a person who is able by imagination, talent or skill to create works of aesthetic value" then I would contend that the professional engineer is very much an artist. Engineers and artists are creative problem solvers. There is obviously a great deal of artistry goes into the streamlined shapes of the sailing vessels, aircraft and motor cars we design as well as the spectacular bridges we have produced over the years. There is little doubt that these creations through their elegance and design elicit emotional responses from the public. I recently sat in on a discussion between my brother, who was, at the time, the general manager of a large computer manufacturing company and an old work colleague whom he had just re-met. Their conversation took them back to the days when they worked together on the design of mechanical calculating machines. The fondness which moved them to talk of these innovative creations had to be seen to be believed. Professor Carl Mitcham, a philosopher of technology at the Colorado School of Mines puts forward the view that: "(An engineer's) invention causes things to come into existence from ideas, makes the world conform to thought, whereas science deriving ideas from observation makes thought conform to existence" *(13)*.

So I considered that introduction of the engineer's practical assessments, supported by his inventiveness and creative artistry, would help to bring a novel approach to my study. Indeed, as you will read in the Appendix to this book, it was largely practical considerations that encouraged me to accept Christian moral values.

The most appropriate definition of faith I have come across is given by Alister McGrath, Professor of Theology, Religion and Culture at King's College in London. In his book *Dawkins' God* *(14)* he provides the definition of faith offered by W. H. Griffith Thomas (1861-1924) an Anglican theologian who was one of Professor McGrath's predecessors in his previous post as Principal of Wycliffe Hall in Oxford. The definition he offers is as follows:-

"(Faith) affects the whole of man's nature, it commences with conviction of the mind based on adequate evidence; it continues in the confidence of the heart or emotions based on conviction, and it is crowned in the consent of the will, by means of which the conviction and confidence are expressed in conduct."

As I have already indicated my main aim in writing this book was to assess whether modern science could help add to my knowledge and help ensure me that I had "adequate evidence" to support belief in a Creator God. I wanted to discover if any of the tremendous developments in science and engineering that had taken place during the last century could influence the evidence I already held for belief in a supernatural power and intelligence responsible for the creation of our universe. I was not convinced that science itself could prove or disprove the reality of a Creator God but I did believe that it could be used to support arguments for and against belief in the supernatural. I particularly hoped that my study would help others, with doubts like me, to see that when it comes to reality, belief in God is essential.

This book describes my search for evidence and what a rewarding search it proved. Initially I was taken aback by how much the science in which I had been schooled had become out of date. However, my search led me through several works which were inspired, enlightening and stimulating to both intellect and imagination. The discoveries of *quantum physics* (I prefer *quantum mechanics* – it gives a better description!) have radically changed my view of ultimate reality and learning from outstanding individuals who have studied both science and religion in depth has proved to be a most enriching experience.

CHAPTER 1

The Project

What has 'theology' said that is the smallest use to anybody? When has' theology' ever said anything that is demonstrably true and is not obvious? What makes you think that 'theology' is a subject at all?

*- **Richard Dawkins** (Emeritus Fellow of New College Oxford)*

....philosophy is dead. Philosophy has not kept up with the modern developments in science, particularly physics.

*- **Stephen Hawking** (Professor of Mathematics, Cambridge University)*

We are not just information processing systems. We are also conscious appreciators of the meaning of information, and creative initiators of new processes of thought.

*- **Keith Ward** (formerly Regius Professor of Divinity at the University of Oxford)*

1.1 Introduction

In his well-known book *The God Delusion (1)* the prominent neo-atheist, Richard Dawkins, describes *the God Hypothesis*[1] as follows:-

"There exists a superhuman, supernatural intelligence who deliberately designed and created the universe and everything in it, including us." Unlike Dawkins I believe in the existence of a creator God of supernatural power and intelligence. I have formed this belief after many years of thought and personal experience and, particularly recently, I have felt well supported by the works of leading thinkers such as John Lennox *(5)*, Alister McGrath *(4)*, John Polkinghorne *(15)* and Keith Ward *(16)*.

I also believe that God exists as an everlasting consciousness. There is something that has thoughts, feelings and perceptions, but no physical body or brain. As a guide, I find it helpful to think of God as an unembodied mind, a pure Spirit that has knowledge and awareness. We are all part of this Mind. I don't accept that the physical universe is the ultimate reality[2] but believe in a model where the ultimate reality has the nature of mind or consciousness.

---

[1] This definition was selected by Richard Dawkins after consideration of a number of alternatives put forward on page 52 of his book *The God Delusion.*

[2] For the purposes of this book I define *reality* as the state of things as they actually exist. *Ultimate reality* is the deepest absolute nature of all things.

I have found that Keith Ward *(17)*, provides reassuring support for this decision. He states that while he considers that the reality of God is infinitely greater than that of any human-like mind, we will not go far wrong if we think of God as a mind, recognising that we are using a model suitable for us, but one that does not exactly apply to God. Much of the material used in the first sections of this chapter was sourced from recent works by Keith Ward, *Why There Almost Certainly is a God (6), The Big Questions in Science and Religion (24)* and *God and the Philosophers (57).* Not only is Keith Ward a cleric, he is also a philosopher, theologian and scholar. He was formerly Regius Professor of Divinity at the University of Oxford. I am grateful for his insights.

I have also been greatly influenced by the works of C. S. Lewis *(18)* and recently I was surprised but pleased to read of the belief in God professed by the brilliant scientist, Max Planck *(19)* who is credited with the discovery of quantum physics. However, like most believers I often have doubts and, in my experience, scientists are increasingly expressing disbelief in God. As indicated in the preface, atheism is being trumpeted by neo-atheists such as Richard Dawkins *(1)*, Daniel Dennett *(20)*, the renowned author Christopher Hitchens *(21)*, and former Professor of Chemistry at the University of Oxford, Peter Atkins *(22).* Even the world's most famous living physicist, Stephen Hawking, has, fairly recently, come out against belief in God *(23)*. So I started writing this book largely because of four main doubts, involving the physical sciences, which were proving deeply troubling to me.

These doubts concerned (1) the strength of the materialist's argument, (2) the influence of evolution on religious belief, (3) the ability of science to explain reality, and (4) the ability of science to support religion. If the materialists were shown to be correct then atheism would easily win the day and certainly my vision of God would be disproved. I was also bothered by Richard Dawkins' claim that evolution supported the view that religious faith is a delusion. If it were the case that the physical sciences in and of themselves, could reliably explain ultimate reality, I would have to reassess my belief in God. Perhaps some of the recent advances in science could assist with these perplexing problems and help me to discover if evidence from the physical sciences could be used to provide positive support for belief in the supernatural.

I would like to stress here that at this stage my main aim was not to determine the moral or physical nature of God. I was simply looking for science to help me find an answer to the question, "Is there a creator God?" where the creator God can be defined as stated by Richard Dawkins at the start of this chapter. The limitations of science[3] make it clear that it would have been unwise to expect much more.

Nevertheless, I do deal with the nature of God but this is left until the Appendix.

---

[3] The limitations of science are considered in some detail in Chapter 4.

1.2 Main Doubts

In the following four subsections I explain my initial doubts. In the first subsection I consider my first doubt which relates to the issues of materialism, and, is specifically concerned with the relationship between mind and matter when trying to explain consciousness. I then go on to the second doubt which deals with evolution and examines the strength of Richard Dawkins' atheist claims for the influence of evolution on religious belief. I then give a description of my third doubt on the reliability of science when trying to explain reality. Explanation of this third doubt also involved an examination of the validity of *personal explanation.* I close by considering my uncertainty on the ability of science to assist belief in the supernatural.

Materialism and the mind/matter relationship

It is generally agreed that the mind is a faculty of human consciousness. We experience consciousness through thought, perception, emotion, will, memory and imagination. When it comes to explaining consciousness, scientists run into a number of difficulties. How do conscious states arise from physical brain states? We seem pretty sure that conscious states can arise from brain states but we do not know what sorts of connections conscious states have with brain states. For instance, how does the conscious experience of 'seeing a motor car', connect with physical brain states and the associated electrochemical activity in the brain? How can it be possible for conscious experience to arise out of the electrochemical activity in the brain? This problem has led to an enormous amount of debate on the relationship between mind and matter, particularly the conflict between *monism* and *dualism.*

In the philosophy of mind, dualism gives the view that mind and body function separately, without interchange. In contrast monism states that mind and body are the same thing. Naturalists and materialists are atheists and believe in monism. However, as someone who believes in God I take a dualist standpoint and when I am asked what my mind is I intuitively identify it with myself, my personality or my soul. I do not believe that my mind is merely my physical brain. A number of philosophers believe that the mind is not completely physical. Rene Descartes *(25)* is reputed to have been among the first to identify mind with consciousness and self-awareness and to distinguish this from the physical aspects of the brain.

The famous Christian apologist C. S. Lewis *(26)* stood strongly against the monism of naturalists and materialists and defended his position by putting forward what has become known as *the Argument from Reason.* Essentially this argument rests on the fundamental difference between material things and non-material mental states. This difference leads us to ask how a neuron[4] can recognise the validity of an argument. How can purposeless physical

---

[4] Neurons are nerve cells that process and transmit information in the brain.

processes lead a thing to intentionally choose for itself? If everything is governed by the rules of science, what then does it mean for someone's mind to be changed by a spaceless, timeless, non-material argument? Lewis contends:- "If, as monism implies, all our thoughts are the effects of physical causes, then we have no reason for assuming that they are also consequent of a reasonable ground. Knowledge, however, is apprehended by reasoning from ground to consequent. Therefore, if monism is correct, there would be no way of knowing this – or anything else – we could not even suppose it, except as a fluke."

Among modern-day philosophers, Keith Ward *(27)* also rejects monism and states:-"Finite minds come into existence when a complex neural network exists. We can formulate a rule that whenever some such neural network exists, then conscious states will exist. But that is a causal statement, not a statement that reduces conscious states to nothing but physical states." He then goes on to argue, "We are not just information processing systems. We are also conscious appreciators of the meaning of information, and creative initiators of new processes of thought."

These arguments seemed pretty convincing to me. However, materialist philosophers, ranging from Thomas Hobbes *(28)* in the 17th century, to modern philosophers such as Daniel Dennett *(20)*, would claim that the intuitions, which led me to my decision about my mind, were misleading. I was particularly taken by Daniel Dennett's book, *Consciousness Explained*. It consists of over 500 pages of tightly packed text, the occasional illustration and much inventive and original thought providing an empirical theory of the mind and dealing with the philosophical problems of consciousness. This book gave me much cause for thought. I had chosen to believe that minds and brains are radically different. I had rejected the materialist's argument, accepted a dualist's position and believed that my mind was not material and existed outside time and space. This made no sense in the terms of the mechanistic science I had been taught.

While struggling with this problem I came across an intriguing book, *The Science Delusion (29)*, by the biochemist Rupert Sheldrake, which argues that my choice is not limited to materialism or dualism and introduces *mental matter.* Conscious matter? I was intrigued and with some enthusiasm, concluded that the argument put forward merited further careful study. I shall deal with this further in Chapter 2. Here, however, I simply wish to point out that my major concern was that, while Sheldrake's third option differed greatly from the standard materialist view, it still centred on matter and this did not align with the dualist stance I had taken. To believe in the God hypothesis I had to reject materialism completely.

I had been raised on Newtonian physics and enthused over Newton's brilliant insight and elegant equations. In their book *The Matter Myth (41)*, scientists Paul Davies and John Gribbin talk of "the Newtonian world view with its doctrine of materialism and the clockwork universe". As an engineer, and a collector of antique timepieces, I was particularly taken with their statement:- "At the time of publication of the Principia *(42)* the most sophisticated machines were clocks, and Newton's image of the working of nature as an elaborate clockwork struck a deep chord. The clock epitomized order, harmony and mathematical precision, ideas that fitted well with the prevailing theology". Not only did these ideas fit in well with the theology of Newton's time they fitted well with my thinking at the start of this project and helped to fire further doubts about my rejection of materialism.

It seemed to me that I had made a rather big decision and I began to think that perhaps I had not given enough thought to the materialist's case. Increasingly doubts were beginning to emerge and I decided that the first aim of any further study should be to investigate the strength of the materialist's argument by looking into the relationship between materialism and reality.

Evolution

My second main doubt concerned the influence of evolution on my thinking. I had been influenced by the claims of Richard Dawkins. When I first considered his work, Dawkins was listed as an emeritus fellow of New College Oxford with atheist and humanist views. He had become known as *Darwin's Rottweiler* for his support of Charles Darwin's evolutionary thinking and in 1976 he wrote, *The Selfish Gene (30).* This work gave what can be considered as a "gene-centred" view of evolution. He followed this in 2006, shaking the world of religion with his highly controversial book,

The God Delusion (1) which quickly sold over 2 million copies. He claimed emphatically that we were created by a process of evolution without any supernatural assistance and also claimed that religious faith is delusive.

Dawkins' work has caused enormous controversy but I believe that, as far as science is concerned, there is little doubt that his claims on the validity of evolution are true. However, what was not clear was how my acceptance of the principles of evolution would affect my belief in a creator God. Dawkins is clear that you cannot accept evolution and believe in a creator God. While I accepted his views on evolution, I was troubled by his aggressive and unbalanced approach to religion. In his book *The Selfish Gene (31),* he states that faith "means blind trust in the absence of evidence, even in the teeth of evidence". He is also quoted as stating, "Faith, being belief that isn't based on evidence, is the principal vice of any religion." While if he had used "blind faith" instead of simply "faith" in these statements, I would have had some sympathy with his claims, I considered that there was, perhaps a lack of balance in his arguments!

In *The God Delusion (32)* Dawkins makes statements like: "The God of the Old Testament is arguably the most unpleasant character in all fiction: jealous and proud of it; a petty, unjust, unforgiving control-freak; a vindictive, bloodthirsty ethnic cleanser; a mysoginistic, homophobic, racist, infanticidal, genocidal, filicidal, pestilential, megalomaniacal, sadomasochistic, capriciously malevolent bully." This ferocious outburst certainly could not get my support. I maintain that, to obtain a balanced and worthwhile view of the Christian God, proper weight must be given to the God of the New Testament of the Christian bible. Here we obtain a clear picture of a loving, caring God who through Jesus' teachings advocates caring and mercy for all. In the New Testament there are many examples which show that God is a God of love. Using William Barclay's translation of the New Testament *(168)*, I have listed five in the footnote below[5].

I am sure Professor Dawkins would have found a solution closer to the truth if his approach had been more balanced and he had simply tackled the question, "Why is the loving and caring God of the New Testament apparently so badly represented in the Old Testament?"

My views on Richard Dawkins' work are shared by Alister McGrath. In his book *The Dawkins*

---

[5] **Nonviolence** – "Then Jesus said to him "Put your sword back in its place.....All who draw the sword, die by the sword." (*Matthew ch26 v52)*. **Peacemakers** – "O the bliss of those who make friends with each other, for they shall be ranked as the sons of God." (*Matthew ch5 v9)*. **Love your enemies** – "I say to those who are listening to me: Love your enemies. Be kind to the people who hate you. Bless those who curse you. Pray for those who abuse you" (*Luke ch6 vs27-28*). **Love your neighbour** – "You must love the Lord your God with your whole heart, and your whole soul and your whole strength and your whole mind and you must love your neighbour as yourself." *(Luke ch10 v27)*. **A new commandment** – "I give you a new commandment – to love each other. As I have loved you. You too must love each other." (*John ch13 v34)*.

Delusion *(4)* he writes[6]: "When I read *The God Delusion* I was both saddened and troubled. How, I wondered, could such a gifted popularizer of the natural sciences, who once had such a passionate concern for the objective analysis of evidence, turn into such an anti-religious propagandist with an apparent disregard for evidence that was not favourable to his case? Why were the natural sciences being so abused in an attempt to advance atheist fundamentalism? I simply cannot understand the astonishing hostility he displays towards religion."

However, while I agreed with Alister McGrath's comments, and was taken aback by the outpourings from Richard Dawkins, the ethologist's claims on religion still troubled me. Dawkins is an exceptionally gifted scientist with an outstanding knowledge of his subject. Could I really hold on to my belief in God and accept the validity of evolution? I decided that I should investigate further.

Scientific and Personal Explanation

There is more than one sort of explanation as to why things happen the way that they do. Of particular importance to this study are scientific explanations and personal explanations and here I consider both sorts.

---

[6] This quote appears on page 10 of the introduction to Alister McGrath's book, *The Dawkins Delusion,* where he expresses his dismay at Richard Dawkins' abuse of the natural sciences in his attempt to advance atheist views.

I first came across personal explanation when, many years ago, as a teenager, I talked with Andrew Douglas, who was then the minister in my local Church of Scotland church. At that time I considered that only common sense and scientific thought could provide satisfactory answers and, in my view, religion was not scientific and fell short of achieving common sense. With some trepidation I put these views to my minister. However, instead of meeting me head on, as I had anticipated, he replied that I could very well be right but before firming up on my views I should consider that there was more than one type of explanation as to why things happen. He continued, "Suppose I am driving along the street in my car when I suddenly think that a child is about to run in front of me. I quickly apply the brakes, bringing the car to rest." He then asked the question, "Why did the car stop?"

He went on to explain there were two possible answers. I could give a scientific explanation, *I might reply that the driver pushed his foot onto the brake pedal causing pressure to be exerted on the fluid in the brake master cylinder then pressurizing the fluid in the pipes which led to the brake mechanisms on the car wheels. This activated the brake wheel discs to clamp the wheels and stop the car.* Or I could give a personal explanation. I would say that he consciously and intentionally stopped the car to avoid hitting a child.

Clearly, scientific explanation deals with physical causes and general laws while personal explanation deals with desires and intentions. In accepting the God hypothesis I had used personal explanation and agreed with Keith Ward's statement *(33)*:- "The God hypothesis connects personal and scientific explanation by postulating that there is an overarching cosmic personal explanation that explains physical states and laws as means to realizing some envisaged purpose."

Not all scientists agree with the validity of personal explanation. It is considered that such explanation belongs essentially to theological philosophy. As an engineer steeped in the use of scientific thought, it is not surprising that I was beginning to have doubts. Like most scientists I was uneasy about mixing philosophy and theology with science. I have found that, in general, scientists are suspicious of philosophers and theologians. In particular, Richard Dawkins *(34)* is scathing in his criticism of theology, stating, "Admittedly, people of a theological bent are often chronically incapable of distinguishing what is true from what they'd like to be true." The famous physicist Stephen Hawking, in his recent book *The Grand Design (35)* states, "...philosophy is dead. Philosophy has not kept up with modern developments in science, particularly physics. Scientists have become the bearers of the torch in our quest for knowledge". These scientists also seem to think that the laws of physics on their own can create reality. Richard Dawkins is quoted as stating *(36)*, "Evolution is the universe's greatest work, and life is arguably the most surprising and most beautiful production that the laws of physics have ever generated." While Stephen Hawking *(37)* claims that all that is needed to create the universe is the law of gravity.

So while theological philosopher Keith Ward considers that the God hypothesis presents us with "a very elegant solution", and, guided by his reasoning, I had chosen personal explanation to help fully explain physical states and I was not yet completely convinced that personal explanation was necessary. I decided that in the next stage of my study I should include an investigation on whether science on its own could produce a better and more reliable explanation of reality than the explanation I had chosen.

Supporting evidence from science

My final doubt concerned the ability of science to support belief in a creator God. In my review of the literature concerning science and religion I had come across numerous scientific papers which appeared to provide clues giving strong supportive evidence to the existence of a creator God. I was particularly impressed by three of them. The first clue involved the wonderfully impressive Human Genome Project led by Dr Francis Collins (7). The second dealt with the "Goldilocks" effect (38) and the third concerned the intelligibility of the universe (39).

The first clue concerns the human genome which consists of all the DNA of our species and provides the hereditary code of life. This strange cryptographic four-letter code forms what Francis Collins calls an instruction book. This was an awesome discovery for me and, at first sight, it certainly looked as if a "designer" has been responsible for our human form. For the second clue the earth also appears to be particularly well suited for life. This can be considered as the "Goldilocks" effect *(38)* where like Goldilocks' porridge in the tale of "Goldilocks and the Three Bears", the universe appears to be "just right" for life. Our distance from the sun provides a good example of this effect. This distance is just right to support life. If it was shorter our atmosphere would prove too hot and if it was longer we would freeze.

So there appeared to be sound evidence of a "higher intelligence" at work in our design and in the design of our universe. If you add to that my third clue which concerns the intelligibility of the universe to us, it certainly appeared that there was something here that should be investigated particularly when we note the statement from Albert Einstein, "The eternal mystery of the world is its comprehensibility" *(39)*. The importance of intelligibility has also been stressed by John Polkinghorne *(40)* who states, "It is intelligibility (rather than objectivity) that is the clue to reality."

Before leaving this section there is a final clue which, I believe, merits attention. This clue came from Francis Collins' book, *The Language of God – A Scientist Presents Evidence for belief (7)*. I was particularly intrigued when, in the first chapter, the author introduced *The Moral Law*. This law had been brought to prominence in the book *Mere Christianity (18)* by the famous Oxford scholar C. S. Lewis who argues that we have within us the sense of right behaviour and character and that "human beings all over the earth have the curious idea that they ought to behave in a certain way and cannot really get rid of it." He goes on to argue further that if the Moral Law exists then there must be a law giver and that law giver is God. Francis Collins relies very heavily on this law to support his belief in God. I was keen to find out more and I report on this later.

At this stage, I will not devote further discussion of questions of morality since, for reasons which I will explain, I do not intend to consider the moral nature of God until the Appendix at the end of this book where *The Moral Law* is critical to my argument.

How best should I investigate the doubts I have described in this chapter? In the true fashion of a research engineer I planned to embark on a project.

1.3 A Project Outline

In Chapters 1-6, I report on this project where my main effort went into researching some of the more recent advances in the physical sciences and investigating their effect on my belief in a creator God.

As already indicated my first priority was to investigate the relationship between materialism and ultimate reality. Chapter 2 describes this investigation which provided a rewarding and enlightening look into the world of *theoretical physics*. I was introduced to the incredible world of *quantum mechanics* and I became fascinated by the range of subjects I encountered, which dealt with a host of different types of *subatomic particles*, an introduction to *string theory*, disappearing *black holes,* mysterious *dark matter* and *dark energy*, *chaos theory* and the *holographic universe*, At the end of this phase of the project I was able to come to some firm, if surprising, conclusions.

Chapter 3 deals with evolution. I examined the strengths and weaknesses of the cases put forward by *evolutionists, fundamentalists and intelligent designers.* The strongest case soon became clear. I then looked at the religious controversy caused by Richard Dawkins and assessed the ability of the evolution argument to fully explain how we got here. Next I considered the question of whether I could accept evolution and believe in a creator God. To help with this I investigated *Theistic Evolution (43)* as proposed by Francis Collins.

Having completed the first two phases of the project I moved on to an investigation of the effectiveness of science in explaining ultimate reality. This investigation is described in Chapter 4. I wanted to determine if science, on its own, could provide a reliable explanation of reality. A look back over the last hundred years or so, showed amazing advances in our scientific knowledge. These advances have greatly changed our perception of what is real and also revealed how difficult it is to predict future changes. I was disappointed to find that substantial disagreement existed between theories covering the physics of the large scale and those dealing with the small scale. I also discovered that we had to be careful with the level of authority we attributed to different areas of science, and it became very noticeable that some recent developments in science were helping to create a culture where use of philosophy and belief in the supernatural was becoming more acceptable to certain scientists.

When I completed this third phase of my project I had formed distinct views of the limits of the physical sciences particularly when explaining ultimate reality and answering the question of the existence of a creator God. I had addressed my first three main doubts.

To deal with my last doubt I began a search with the aim of finding, in the physical sciences, examples of instances where science could produce evidence to support belief in the supernatural. I called these clues "Signposts from science to reality". This proved to be a reassuring exercise and I found several examples. In Chapter 5 I have given particular attention to the three clues I mentioned earlier in this chapter. These clues involve the intelligibility of the universe, the 'Goldilocks Effect' and finally the famous 'Human Genome Project'.

I finished my project by conducting a review of my findings and in Chapter 6 I report on a study which had proved to be surprising, exciting, game-changing and extremely worthwhile. The study has proved pivotal in altering my views on the ultimate reality of our universe. I love learning and I had learned much. I was able to reach, with some confidence, a decision on whether the additional information I had gained from my project had finally helped me to find the 'adequate evidence' I was searching for.

1.4 Key Points

For this chapter the main point to be noted is that the central aim of my project was to strengthen my faith in a creator God by employing science to provide answers to the following questions: -

- How credible is the philosophy of materialism?
- Does acceptance of the theory of evolution negate belief in God?
- Is science fully equipped to answer the *God Question?*
- Can science produce evidence to support belief in the supernatural?

The moral nature of God was not considered in the main project.

CHAPTER 2

Materialism and Reality

Most people have rejected scientific values because they regard materialism as a sterile and bleak philosophy which reduces human beings to automatons and leaves no room for free will or creativity. These people can take heart: materialism is dead.

*- **Paul Davies and John Gribbin** (Science authors)*

Everything we call real is made of things that cannot be regarded as real.

If quantum mechanics hasn't profoundly shocked you, you haven't understood it yet.

*- **Neils Bohr** (Quantum physicist)*

Fully 70% of the mass density in the universe appears to be in the form of dark energy. Twenty six percent is dark matter. Only 4% is ordinary matter. So less than 1 part in 20 is made out of matter we have observed experimentally or described in the standard model of particle physics.

  - Lee Smolin (Theoretical physicist)

2.1 Introduction

Theists, like me, believe in a reality they do not fully understand while the reality of the materialist is well understood, and when I started to write this book I considered that, materialism, one of the big weapons in the atheist's armoury, was a strong candidate for providing one of the most obvious, reliable and simple explanations of reality. It seemed that materialists could convincingly adopt atheist views and readily dismiss the supernatural. They argue that our universe contains only solids, liquids, gases and an increasing number of other sorts of matter. We encounter physical objects every day. Our senses can directly perceive when material things are present and when we cannot sense them immediately we have devised instruments to do the detecting for us. Since the 17th century until today, materialists like Thomas Hobbes **(28)**, Daniel Dennett **(20)** and Richard Dawkins **(1)** have put forward some very strong arguments in support of their views. When we give priority to observable reality for justifiable reasons, materialism provides an obvious explanation for the way things are. So in my search to find a true explanation of reality an assessment of the reliability of the materialist's argument was an early priority.

First, I consulted the *Oxford Dictionary **(44)**,*

which gave the following definition for materialism:- "Opinion that nothing exists except matter and its movements and modifications also that consciousness and will are wholly due to a material agency." I was a little perplexed when I first read this, however, further investigation led me to understand that there are two types of materialist. The opinion expressed in the first part of the above definition – "that nothing exists except matter and its movements" – is held by *radical materialists*. In the complete definition the scope has been amended to accommodate the views of *emergent materialists* who believe that minds, and hence consciousness and will, somehow emerge from matter.

Next I decided to seek a definition of matter which would satisfy the materialist. Finding a satisfactory definition was more difficult than I had imagined. After browsing the internet I came to the conclusion that the term "matter" is used throughout physics in a large variety of contexts: there is, for example, "elementary matter", "condensed matter", "strange matter" ,"anti-matter", "dark matter", and "nuclear matter". I came to the conclusion that, in physics, there is no broad consensus on a definition of matter, and the term "matter" is usually used in conjunction with some modifier. Finally I decided to settle, at least temporarily, for the simple definition:- *"Any substance which has mass and occupies space" (45)*.

As a first step in determining the validity of the materialist's view of reality I decided to turn to the physical sciences. I was intrigued by all of the exciting facts that were emerging from the field of theoretical physics and I considered that I should look a bit closer at the significance of some of the more pertinent recent developments. Clearly, since this was not my subject area I could only hope to get an initial grasp of the complex details. However, I thought my knowledge of thermodynamics would prove useful and, although limited, an overall appreciation of the facts and some sense of their significance, was possible. My brief look into the field proved to be very rewarding and indeed, fascinating. I learned much.

In this chapter I consider the sciences associated with materialism and I begin with a look at the major changes which have occurred in these sciences due to advances made in the last century. I then investigate the nature of matter and present some of the most recent exciting advances which challenge the veracity of current views of the material world. First, I look inwards to the world of subatomic particles where *Quantum Theory*, *String Theory* and *Chaos Theory* prove to be particularly relevant to my study. Then I take a look outwards into the cosmos where the fairly recent discovery of *Dark Matter* and more importantly *Dark Energy* are introducing evidence that is greatly changing the way in which we perceive the reality of the universe. In the final section I put forward my conclusions which include my assessment of the claim that the material world presents ultimate reality.

2.2 Modern Physical Sciences

It was with considerable disappointment that I discovered that the Newtonian physics of which I was aware, and had been taught as a student in the 1960s, was limited. I had been raised on Newtonian physics and never ceased to be delighted at the way Newton's elegant equations could be used accurately to describe the physical world around me. As a mechanical engineer I found that Newton's work was of particular importance. He has given us the laws of mechanics and, on countless occasions, I have been more than grateful to employ his laws. In the realm of engineering science, Newton was, without doubt, "the king". However, things have now changed immensely in the field of engineering. An example of these changes has already been mentioned in the preface, where I have recorded that, even at the Massachusetts Institute of Technology, one of the truly outstanding researchers in the field of *quantum physics,* Seth Lloyd *(9)*, is a professor of mechanical engineering.

As a rough guide we can say that nowadays the disciplines of physical science are divided into *Newtonian physics* and *Modern physics.* Essentially Newtonian physics can be used as the physics of everyday objects. Modern physics describes the less familiar world we observe when we go beyond the everyday. The age of modern physics was started towards the end of the 19th and early 20th centuries when experiments were performed which could not be explained by the Newtonian laws. One part of modern physics, *Relativity,* which was introduced in 1905 by Albert Einstein *(46)* and is much used in the field of *Cosmology*, deals with conditions where objects that are moving at very high velocities or are in the presence of strong gravitational forces. Under these conditions, relativity predicts that moving clocks tick more slowly than an observer's stationary clock and moving objects are shortened in the direction that they are moving with respect to the observer. Measurements can be made to confirm these predictions.

The other part of modern physics is quantum physics, which deals with light and things that are very small such as molecules, atoms and subatomic particles. It is used in the field of particle physics and, as we shall see later in this chapter it can be used to describe a strange world where objects don't exist until they are measured and particles can move in and out of existence.

It is important to note that, while we can still use Newtonian physics, this class of physics is limited. Newtonian theory proposes that all material objects are made up of particles which exhibit only particle properties. Quantum theory proposes that matter is formed from quantum particles which exhibit both particle and wave properties. At the extremely small scale of subatomic particles the wavelengths encountered are comparatively large enough to influence how the quantum particles operate but as the mass of a particle increases its wavelength gets shorter and shorter. For the everyday objects we observe the wavelengths encountered due to quantum effects are negligible and their influence can be ignored. This means that, to my relief, we can still continue to use Newton's theories but within limits and we must be careful to remember that all material objects are made up of quantum particles.

2.3 The Nature of Matter

2.3.1 Looking inwards

To determine the nature of matter I started by looking inwards towards the strange world of particle physics. This investigation led me to the incredible predictions given by *quantum theory, string theory and chaos theory.* I consider these theories in the next three subsections.

Quantum weirdness

I was astounded at how much had changed since I had last visited the area of particle physics. When I was a student I was taught that *matter* was composed of atoms formed from the subatomic particles *protons, neutrons and electrons.* I now found a very different situation. Protons, neutrons and electrons are still there but the proton and neutron are not fundamental particles and the electron is one of a family of fundamental particles called *leptons.* Protons and neutrons each consist of three smaller particles called *quarks.* The quarks come in two varieties named *up* and *down*. *A* proton consists of two up-quarks and a down-quark and a neutron consists of two down-quarks and an up-quark. All of the matter in our physical world appears to be made from combinations of leptons, up-quarks and down-quarks. In the mid-1950s conclusive evidence was found for the existence of another fundamental particle, the *neutrino*. I was fascinated by the existence of this "ghostly" particle that can pass through many trillion miles of lead without the slightest effect on its motion.

The discovery of additional *particles* did not end with the neutrino. Many more have been reported. For those not involved in this particular scientific area it seems that, as time passes, there is a fairly constant flow of new particles being discovered in particle accelerators such as Fermilab in Illinois in the USA *(47)* or the Hadron Collider at CERN in Geneva in Switzerland *(48)*. The situation seems to be very complicated and I have some sympathy with Keith Ward *(49)* when he states: - "It no longer seems to be a set of simple elementary particles. Instead we have a *particle zoo* of flickering insubstantial virtual wave particles."

However, particle physicists form an extremely bright and resourceful group, and, to their immense credit, they have managed to organise things for known matter. At the present time they have reduced the number of standard fundamental particles to sixteen. There are six quarks, six leptons and, just when you think things are becoming simpler, four rather complex "particles" called *bosons*. Bosons which are not "matter" are often referred to as *force particles* providing interaction energy inherent to composite particles and contributing to the mass of ordinary matter. Clearly, despite these heroic efforts at simplification the situation remains very complex and several vital questions remain to be answered, such as, "Why are there so many fundamental particles?" and, "What drives their number?"

Physicists have now completed the *Standard Model* of particle physics. This model is claimed to be almost a complete theory of the fundamental interactions which mediate the dynamics of known subatomic particles. To complete the verification of the theory scientists have had to confirm the probability of the existence of a particle called the *Higgs boson* and they have only recently obtained experimental data in the particle accelerator at CERN in Geneva. Clearly the model is the result of some truly outstanding research and represents a major advance in the field. Even a brief assessment of the complexities of quantum physics will give you some idea of the tremendous result. While the casual observer might think that, in view of this achievement, the position of the materialist is strengthened, I was about to find that the materialists' contentions were both naive and untrue. In my project, I had now reached the strange and completely non-intuitive realm of quantum theory.

This theory is now the basis of modern physics as it explains the nature and the behaviour of matter and energy at the atomic and subatomic level. It was initiated around 1900 when the famous scientist Max Planck introduced the assumption that light was not a continuous wave but made of individual units or *quanta (50)*. This work was followed in 1905 by the research of Albert Einstein who used Planck's work to explain the *photoelectric effect (51)*. Some 19 years later in 1924 a young French aristocrat, Louis de Broglie, proposed that there was no fundamental difference in the composition or behaviour of energy and matter. He claimed that elementary particles of both energy and matter behaved, depending on conditions, like either particles or waves. As a reward for his efforts he won the Nobel Prize for Physics in 1929. In 1927 the German physics theorist Werner Heisenberg finally shook the establishment with his famous *Uncertainty Principle,* which proposed that precise simultaneous measurement of two complementary values – such as the position and momentum of a subatomic particle – is impossible. A further direct and concise explanation of this principle may be found in reference *(52)*.

Quantum theory[7] arose with the discovery that subatomic particles are discrete packets of energy

---

[7] For an introduction to quantum mechanics I can recommend, *Quantum Theory – A Very Short Introduction,* by John Polkinghorne *(53)* and *How to Teach Quantum Physics to Your Dog (54)* by Chad Orzel. The eminent and distinguished scientist Sir John Polkinghorne is a Fellow (and former President) of Queens College, Cambridge and he has produced an incisive and lucid study which I found exceptionally useful. I make several references to his work in this book. Chad Orzel is a Professor in the Department of Physics and Astronomy at Union College in New York and not only is his book wonderfully informative, it provided me with a most amusing and entertaining read of a scientific text. The thoughts of Emmy, the dog, are used with pleasing and original effect to illustrate the clear explanations provided.

with wave-like properties and it is important to be clear that when quantum physicists refer to a "particle" they do not mean particle in the common sense of the term. Quantum particles, such as the photon, which form light, exhibit *wave/particle duality* and they behave in a way resembling both particles and waves. Also Heisenberg's uncertainty principle showed that analysing quantum particles required a statistical approach. Quantum particles have uncertain boundaries and their properties are known only as possibilities. The nature of their interactions with other quantum particles are still a bit of a mystery.

The opening chapter to Chad Orzel's book, *How to Teach Quantum Physics to Your Dog, (55)* makes the following statement:-

"Quantum theory's effect on science goes beyond the merely practical – it forces physicists to grapple with issues of philosophy. Quantum physics places limits on what we can know about the universe and the properties of objects in it. Quantum mechanics even changes our understanding of what it means to make a measurement. It requires a complete rethinking of the nature of reality at the most fundamental level."

It may be surprising to discover that physicists are turning to philosophy to assist them. In Newtonian physics there is an immediate intuitive connection between the theory and the reality we observe but in quantum physics the results are often counter-intuitive. They have to be interpreted without any assistance from intuition and the implications of the interpretations give rise to philosophical issues which deal with the nature of reality.

To help explain existing experimental data there are at present numerous interpretations of quantum mechanics from which to choose (John Polkinghorne *(56)* lists and assesses five), but I believe the main candidates are the *Copenhagen Interpretation,* on which there seems to be several options, and the *Many Worlds Theory.* The Copenhagen Interpretation *(57)* was proposed by the Danish physicist Neils Bohr along with Werner Heisenberg and it asserts that a particle is whatever it is measured to be (for example, a wave or a particle) but it cannot be assumed to have specific properties, or even exist, until it is measured. This translates to the principle of *superposition* that states while we do not know what the state of any object is, it is actually in all states simultaneously, so long as we don't look to check. Therefore, essentially, Bohr was asserting that objective reality does not exist.

The other major interpretation of quantum theory is the *many worlds (or multiverse)* theory. A useful explanation of this interpretation may be found in reference *(58).* It claims that as soon as a potential exists for any object to be in any state, the universe of that object transmutes into a series of parallel universes equal to the number of possible states in which that object can exist, with each universe containing a unique single possible state. Also, there is a mechanism for interaction between these universes that somehow permits all states to be accessible in some way and for all possible states to be affected in some manner.

The reality we observe every day is not ultimate reality and for that final actuality we must look towards the quantum world. Clearly, there is not an obvious explanation of the nature of ultimate reality. If you have not come across quantum physics before I am sure that, like me, you will take some time to appreciate the astounding significance of the above paragraphs which, I suggest, you will view initially with disbelief. It needs to be remembered, however, that quantum theory has been tested to an amazingly high level of precision, making it the most accurately tested theory in the history of scientific theories. You can also take some comfort from the words of Neils Bohr, one of the earliest and most significant quantum physicists, by heeding his words *(59)*, "Everything we call real, is made of things that cannot be regarded as real. If quantum mechanics hasn't profoundly shocked you, you haven't understood it yet."

We should not leave this look into quantum physics without mentioning *virtual particles.* The clearest definition I have found for these particles is given in reference *(60)*, as follows: -

"A virtual particle is an elementary particle of transitory existence that does not appear as a free particle in a particular situation but that can transmit a force from one particle to another."

This is presented along with the following helpful explanation of a virtual particle: -

"A virtual particle is a short-lived subatomic particle whose existence briefly violates the principle of conservation of energy. The *uncertainty principle* of quantum mechanics allows violations of conservation of energy for short periods meaning that even a physical system with zero energy can spontaneously produce energetic particles."

We can conclude that quantum physics undermines materialism largely because it shows that matter lacks the "substance" necessary for the materialists' claim. Also, it clearly shows that ultimate reality is certainly not to be found in the claims of the materialist which give a picture which is much too simple. Research reported in publications such as *Quantum Theory* by John Polkinghorne *(53)*, has shown that the quantum world is an utterly bizarre world where nothing is certain, objects don't have definite properties until you measure them and, as we have just noted, particles can pop into and out of existence. It is not a world of facts but a world of potentialities or possibilities.

However strange this may seem extensive experimental data show that it is real and in quantum reality there is no place for the tangible, solid and reliable reality that materialism claims to represent.

Vibrating strings of energy

Another prediction to influence the nature of ultimate reality is given by *String Theory* which predicts that the fundamental particles of matter (quarks, electrons etc.) are composed of extremely tiny vibrating filaments (and membranes) of energy. I first read about string theory when I came across a book titled, *The Elegant Universe: Superstrings, Hidden Dimensions and the Quest for the Ultimate Theory (61).* In this book Brian Greene, Professor of Physics and Mathematics at Colombia and Cornell Universities, asks his readers to consider a universe where all matter is generated by incredibly tiny loops of energy vibrating in eleven dimensions. These strings form the fundamental building blocks of matter. At this stage in my study it appeared to me that, if string theory was verified, instead of the materialists' assertion that "only matter exists", we could now claim that nothing but energy exists. I wondered if materialists had reacted to this claim. Since it is generally agreed that initially matter was formed from energy, I considered that the statement that *nothing exists but energy* might prove a more valid claim than the one given by the materialist. I consulted the Oxford Dictionary of Philosophy *(62)* which confirmed that philosophers are also unhappy with materialism and: -

"...now tend to prefer the term *physicalism* (the doctrine that the real world consists simply of the physical world) since physics has shown that matter itself resolves into forces and energy, and is just one amongst other physically respectable denizens of the universe." Unfortunately for them, physicalists, who align themselves with this definition, will now have to concede that there are major questions to which they need to respond. Their doctrine does not fit into any of the interpretations of quantum physics and I would suggest that, as a start, they read Chapter 3 of *Miracles (26)* by C. S. Lewis. The material in this chapter titled "The Cardinal Difficulty of

Naturalism" will give them food for thought.

At present, development of string theory is encountering serious problems and, as reported by physicist and mathematician Andrew Zimmerman Jones, in an article listed under *The Nature of Reality* and entitled *Can String Theory be Tested? (63)*[8], confirming by experiment or refuting some of the facets of string theory might not be possible. Nevertheless I would contend that modern physics is emphatically directing us away from materialism.

The Theory of Chaos

In their book *The Matter Myth (41)*, two eminent science research authors, Paul Davies and John Gribbin[9], give strong support to the view that materialism is not valid. Both authors have written several best-selling books on science. Paul Davies is

---

[8] **Andrew Zimmerman Jones** is the author of *String Theory for Dummies (63)* and many wrestling with the complexities of string theory are grateful for his efforts

[9] **Paul Davies** is a popular and well respected physicist and broadcaster. He is also a writer of a series of scientific books which I can thoroughly recommend. He was an outstanding student at University College London where he gained his PhD and he is currently a professor at Arizona State University, having previously held posts at the University of Cambridge, the University of Newcastle upon Tyne and the University of Adelaide in Australia. As might be expected he has keen interests in cosmology and quantum field theory - **John Gribbin** is an astrophysicist. He is currently a visiting fellow in astronomy at the University of Sussex. He writes on a range of subjects including, quantum physics, human evolution and climate change. He also writes science fiction and he has completed some children's books on science. Interestingly both Davies and Gribbin worked for a time under the guidance of the celebrated English astronomer **Sir Fred Hoyle (114)**.

Director of the Beyond Centre at Arizona State University. John Gribbin trained as an astrophysicist at the University of Cambridge and moved to become a Visiting Fellow in Astronomy at the University of Sussex. In a chapter titled "The Death of Materialism" *(65)* they express the view that quantum physics undermines materialism because it reveals that matter has far less "substance" than we might believe. Interestingly they claim that two major developments of the 20th century have "laid to rest" the idea of Newton's material and clockwork universe. The first is quantum mechanics and the second concerns *chaos theory.* The *theory of chaos* is now used in the analysis of dynamical systems and Davies and Gribbin contend that this development goes even further than quantum physics in demolishing materialism.

As an introduction to the subject of chaos I found reference *(65)* helpful. Chaos is the science of surprises of the nonlinear and the unpredictable. It teaches us to expect the unexpected. A chaotic system is a dynamical system that is highly sensitive to initial conditions. In a chaotic system the key feature concerns the way that *predictive errors* evolve with time. In many text books a single simple pendulum is used to illustrate a non-chaotic system and for a chaotic system, a double pendulum (a single pendulum with another pendulum attached to its free end) is used as a demonstration model. The simple pendulum, once started, establishes a regular and predictable motion. Any predictive error, due to errors in setting up initial conditions, will be small and increase only slowly. However, for the chaotic system, once the pendulums are set in motion any small difference between the two identical systems grows rapidly. The motions of the two systems diverge exponentially fast and for the predictive problem any input error increases at an escalating rate. Very soon the error engulfs the calculation and any means of prediction is gone. Small differences in initial conditions (such as those due to rounding in numerical compilations) can yield wildly diverging outcomes making long-term prediction impossible.

A particularly surprising feature of a chaotic system is that it is deterministic. In a deterministic system future states are dictated, through some law of dynamics, by preceding states. Davies and Griffin *(66)*, point out that it used to be believed that determinism went hand in hand with predictability but the chaotic pendulum shows that this is not necessarily the case. For the chaotic pendulum determinism implies predictability only in the idealised limit of infinite precision in setting the initial conditions. Davies and Gribbin *(67)* show that infinite precision is impossible and they state that they can conclude that deterministic chaos seems random because we are *necessarily* ignorant of the ultrafine detail of just a few degrees of freedom. They then make the reader sit up when they add "and so is the universe itself!" They then go on to contend that it seems then that the universe is incapable of computing the future behaviour of even a small part of itself, let alone all of itself. Finally they emphasise the profoundness of this conclusion and stress that even accepting a strictly deterministic account of nature, the future states of the universe "are in some sense open."

Davies and Gribbin then continue to discuss *linear* and *nonlinear systems* and deal at some length with the surprising results from recent investigations of nonlinear systems. They define a linear system as one in which the whole is equal to the sum of its parts and in which the sum of a collection of causes produces a corresponding sum of effects. In contrast to a linear system, the output of a nonlinear system is not directly proportional to the input and nonlinear systems, which generally must be understood in totality, are much more difficult to handle than linear systems. To date most research has been carried out using linear systems despite the fact real systems usually turn out to be nonlinear at some level. However, largely due to the strength of modern computing, investigation of nonlinear systems and processes is now readily possible and Davie and Griffin come up with some amazing examples.

In a section headed "Waves with a will of their own" they consider, at some length, the *soliton* which can be defined as a self-reinforcing solitary wave (a wave packet or pulse) that maintains its shape while it travels at constant speed. It was pleasing for me to note that the first person to record the existence of this type of solitary wave was a Scottish engineer by the name of John Scott Russell in 1834. When he was out riding by a canal near Edinburgh he noted that when a boat, being drawn along the narrow canal by two horses, came to a sharp stop, a large mound of water suddenly appeared at the bow of the boat. When describing what happened later late he wrote that the wave "rolled forward with great velocity assuming the form of a large solitary elevation, a rounded smooth and well defined heap of water which continued on its course along the channel apparently without change of form or diminution of speed". It is reported that Russell followed this wave for two miles before he lost it in a winding channel.

The main physical details to note about solitons are, first, that they are of permanent form. Second, they are localised within a region and third, they can interact with other solitons and then emerge from the collision unaltered except for a phase change. Unsurprisingly they could not be explained until nonlinear systems and modern computing came to the rescue. The soliton is not just a hydraulic phenomenon, they can be found in subject areas of fibre optics, superconductors, molecular biology and even cosmology.

It is worth stressing that most real systems will turn out to be nonlinear at some level and it is informative to turn again to Davies and Griffin *(68)*. Here they comment on the exceptional growth of nonlinear science and put this down to the increasing availability of fast computers. This expanding study of nonlinear systems is causing a remarkable shift of emphasis away from inert "things" to "systems" that contain elements of spontaneity and surprise. In many cases the same basic nonlinear phenomena are appearing in systems that are not really material including computer networks and economic models. So with the machine analogy now looking distinctly strained, the link with Newtonian materialism is fading fast. Davies and Gribbin conclude that "the very breadth of the nonlinear revolution is leading to the rapid demise of the Newtonian paradigm[10] as the basis for our understanding of reality."

So chaos theory shows that Newton's solid and elegant description of a reality consisting of a clockwork universe filled with "things", made from inert matter acted on by impressed forces, has been replaced by a universe filled with "systems" that contain elements of spontaneity and surprise. As claimed in *The Matter Myth (41)* "the old vocabulary of science is giving way to language more reminiscent of biology than physics – adaptation, coherence, organisation and so on". In their book, Paul Davies and John Gribbin present a set of impressive discoveries that challenge our understanding of physical reality. They argue that the paradigm of the universe as a "mechanism" must be replaced and they give convincing reasons for this. The aim of their work is to: -

"….provide a glimpse of the new universe that is emerging. It is a picture still tantalisingly incomplete, yet compelling enough from what can already be discerned. We have no doubt that the revolution which we are immensely privileged and

---

[10] *Mechanical Universe* – The theory of the Mechanical Universe supports Newtonian physics and contends that the universe is best understood as a system composed entirely of matter in motion under a complete and regular system of laws.

fortunate to be witnessing at first hand will forever alter humankind's view of the universe."

I was stimulated and thrilled when I read this statement from two such well respected and reliable scientists.

So, to answer the question posed at the start of this section we can reliably state that the material world does not present us with ultimate reality. We can agree with Davies and Griffin and claim that in the light of the information we have obtained, "The rigid determinism of Newton's clockwork Universe evaporates, to be replaced by a world in which the future is open, in which matter escapes its lumpen limitations and acquires an element of creativity."

In Section 1.2 of Chapter 1 I mentioned that in his book *The Science Delusion (29)* the eminent biochemist Rupert Sheldrake supports and develops this claim. Rupert Sheldrake takes the evidence for his argument from philosophers such as Baruch Spinoza *(69)* and Gottfried Leibniz *(70)*. Relying on the work of Galen Strawson *(72)*, he argues for the validity of *panpsychism* and contends that even atoms and molecules have a primitive kind of mentality or experience. In his argument he makes the distinction between aggregates of matter, like tables and rocks, which are shaped by external forces, and *self-organising systems* like atoms and cells which he states are "complex forms of experience emerging spontaneously". These systems are at the same time physical (non-experiential) and experiential: in other words they have experiences.

At the end of a chapter titled "Is nature mechanical?"*(73)* he summarises his conclusions as follows: -

"The mechanistic theory is based on a metaphor of the machine. But it is only a metaphor. Living organisms provide metaphors for organised systems at all levels of complexity, including molecules, plants and societies of animals, all of which are organised in a series of inclusive levels in which the whole at each level is more than the sum of the parts, which are themselves wholes at a lower level. Even the most ardent defenders of the mechanistic theory smuggle purposive organising principles into living organisms in the form of selfish genes or genetic programs. In the light of the Big Bang theory, the entire universe is more like a growing, developing organism than a machine slowly running out of steam."

Scientists, such as Davies, Gribbin and Sheldrake, are putting together a very exciting picture of a universe which is alive and our outdated vision of an inanimate, "mechanical" universe can now be put to rest.

2.3.2 Looking upwards

Having looked inwards towards the centre of matter, I then turned my attention to looking upwards towards the cosmos. Here, again, I found some startling revelations.

Dark mystery

Only fairly recently has *dark matter*, which we cannot detect directly, been brought to our attention. It can be defined as: - "Matter of unknown composition that does not emit or reflect enough electromagnetic radiation to be observed directly but whose presence can be inferred from gravitational effects on visible matter." It is believed that dark matter plays an essential role in shaping the universe since it provides most of the gravitational pull needed to grow galaxies. Scientists have managed to divide candidates for dark matter into two broad categories, MACHOs (massive compact halo objects) and WIMPs (weakly interacting massive particles). However, they still do not know its composition. As with string theory experimental verification of the composition and properties of dark matter is fraught with difficulties.

Turning now to *dark energy,* the foundations of cosmological theory were rocked in the mid-1990s when two groups of astronomers announced that the expansion rate of the universe was speeding up. Until then cosmologists considered that gravitation was acting as a brake on expansion of the universe, slowing it down from the explosive start at the big bang to the modest rate now observed. However, a mysterious antigravity force opposing gravity had succeeded in transforming deceleration into acceleration. Dark energy was the name given to this anti-gravitating influence. The importance of dark energy cannot be understated and I deal with this further in Chapters 4 and 5. It has a vital effect on the future of the universe. It seems that dark energy contributes most of the mass of the universe but, like dark matter, nobody yet knows what it is.

The distinguished theoretical physicist Dr Lee Smolin states in his book *The Trouble with Physics (74)*:- "Fully 70% of the mass density in the universe appears to be in the form of dark energy. Twenty six percent is dark matter. Only 4% is ordinary matter. So less than 1 part in 20 is made out of matter we have observed experimentally or described in the standard model of particle physics."

2.4 Discussion

What did I learn from this initial study on materialism and reality? How would it influence my thinking for the overall assessment I was undertaking? I had learned a great deal and was much enthused by what I had read. I had uncovered several areas where I was keen to learn more but I felt I had learned enough to now to postpone further reading in this topic area and press on with the rest of my assessment. I began this stage of my study by simply looking for flaws in the materialist's views on reality and ended up having to completely rethink the nature of reality at the most fundamental level.

My study has shown me that the material reality is not ultimate reality. Quantum theory shows that ultimate reality is something much more complicated and much less substantial. Everything in the universe we perceive as physical is built from quantum particles. In modern physics solid particles have been replaced by quantum particles which are neither just waves or just particles but exhibit some wave properties and some particle properties at the same time. As indicated by John Polkingthorne *(75)*, "Quantum reality is cloudy and fitful in its character". The quantum world is an utterly bizarre world where nothing is certain and objects don't have definite properties until you measure them. It is not a world of facts but a world of potentialities or possibilities. Physicists like Davies and Gribbin *(76)* contend that quantum physics undermines materialism and most people have rejected scientific values because they see materialism as a sterile and bleak philosophy reducing human beings to automatons and leaving no room for free will or creativity. Then emphatically they add, "These people can take heart: materialism is dead.*"

The view of the universe as something that looks more akin to what we normally associate with the "biological" has also been put forward by Rupert Sheldrake in his book, *The Science Delusion.* Further supporting this view, in a chapter of his book headed, "Is Matter Unconscious?" *(77)*, he has presented further results of a fascinating study from which he returns to the dualism versus monism argument I considered in Chapter 1. He reports that nobody can satisfactorily explain how non-physical minds can interact with material brains. Materialists have rejected the existence of immaterial minds leaving only unconscious matter. However, since human beings are conscious, this elimination of minds has created a big problem for materialists. As a result they have tried to explain human consciousness away or dismiss it as illusory. Sheldrake asserts that "instead of assuming that materialism and dualism are the only options, some philosophers he advances the idea that all self-organising material systems have a mental as well as a physical aspect."

So Davies and Griffin maintain that matter has acquired an "element of creativity" and Sheldrake also suggests that the consciousness of self-organising material systems is worthy of serious consideration. All three scientists believe that the metaphor of the universe as a biological system is appropriate.

In efforts to explain physical reality, theoretical physicists such as Brian Greene *(61)*, have turned to concepts such as string theory. However, despite some brilliant research, problems have still to be resolved. I believe that the exceptionally able and resourceful scientists concerned with this work will eventually achieve success. However, even if they do not, I would still contend that the results of current research are drastically changing our views on physical universe where ordinary matter contributes so little. We do not yet know much about dark matter and know virtually nothing at all about dark energy. At present we believe that 70% of the mass density of the universe is dark energy while some 26% is dark matter which is so lacking in substance that we can hardly detect it. Only the remaining 4% is matter we know about.

At the beginning of this book I recorded that I rejected materialism. I claimed that the ultimate reality was not the physical universe but had the nature of Mind and I argued for the use of a metaphor in which God, the ultimate reality, can be considered as a Mind. From the study on the science of materialism reported here I have discovered nothing that would cause me to change my belief. Indeed, from information gathered on, (a) the ultimate reality of the quantum world, (b) the signs of consciousness shown by matter formed into systems, and (c) the likelihood of an animate universe seen as essentially an infinite cloud of energy, it seems to me that science is now providing physical clues which make my argument for a metaphysical problem increasingly acceptable. It is clearly showing that the contents of universe, including us, are not of "substance" and it has certainly destroyed the case for materialism.

2.5 Key Points

For this chapter the main points to be noted are as follows: -

- Quantum physics undermines materialism.

- Recent research using chaos theory is tending to show that the reality of a "clockwork" Newtonian universe filled with "things" has been replaced by a universe filled with "systems".

- Matter can be formed into systems which produce elements of spontaneity.

- It is claimed that self-organising systems of matter can show signs of consciousness.

- Consideration should be given to the claim that the metaphor of the universe as a biological system is appropriate.

- Our universe has now been shown to have a mass density composition of 70% dark energy, which permeates all of space, 26% dark matter, which is so lacking in substance that we can hardly detect it, and only 4% of ordinary matter.

- At present we know very little about the composition of dark matter and virtually nothing about dark energy.

CHAPTER 3

Evolution and Reality

Evolution is the universe's greatest work, and life is arguably the most surprising and most beautiful production that the laws of physics have ever generated

*- **Richard Dawkins** (Emeritus Fellow of New College Oxford)*

The shelves of many evangelicals are full of books that point out the flaws in evolution, discuss it only as a theory, and almost imply there's a conspiracy here to avoid the fact that evolution is actually flawed. All of those books, unfortunately, are based upon conclusions no reasonable biologist would now accept.

*- **Francis S. Collins** (Director of the Human Genome Project)*

We do not need an intelligent creator. Blind and gradual selection will do the trick. But the chances of doing so seem to be astronomically small. Unless, that is, that the laws governing the sorts of mutation that occur have been carefully worked out beforehand.

*- **Keith Ward** (Formerly Regius Professor of Divinity at the University of Oxford)*

3.1 Introduction

As indicated in Chapter 1, my initial reason for undertaking this investigation into *evolution* was mainly to reach a conclusion on whether I could believe in the reality of a creator God and also accept the validity of the *theory of evolution.* However, my investigation proved to be more informative than I had originally envisaged and as a result I was able to develop answers to the following four main questions:

(1) What theory gives the best explanation of how humans have developed until now?

(2) Does belief in a creator God prevent acceptance of the theory of evolution?

(3) Does evolution need God?

(4) How valid is *Theistic Evolution*?

In this chapter, I begin by discussing Darwinian evolution[11]. Then, I review what I consider to be the

---

[11] The central idea of biological evolution is that all life on Earth

three most widely accepted explanations of how humanity has reached the present stage in our development. I assess each explanation to decide which process I think provides the best available scientific solution and offer my reasons for selecting my preferred option. I then ask if belief in a creator God prevents acceptance of this option. I follow this with a discussion on the limits of science

---

shares a common ancestor, just as you and your cousins share a common grandmother. The word 'evolution' can have various meanings and, for those not familiar with the evolutionary process considered here, this can cause much confusion. So for this introduction I decided that I should present an explanation of what I mean when, throughout this chapter, I use the word 'evolution', without adding any qualification. A search through information sources on the internet showed the most apt explanation came from experts at the University of Berkeley in California, USA (**78**). Firstly they give the definition as: - "Biological evolution, simply put, is descent with modification. This definition encompasses small-scale evolution changes in gene frequency in a population from one generation to the next) and large-scale evolution (the descent of different species from a common ancestor over many generations)). Evolution helps us to understand the history of life."

This is followed by a short explanation:- "Biological evolution is not simply a matter of change over time. Lots of things change over time: trees lose their leaves, mountain ranges rise and erode but they are not examples of biological evolution because they don't involve descent through genetic inheritance. Through the process of descent and modification, the common ancestor of life on Earth gave rise to the fantastic diversity that we see documented in the fossil record and around us today." When the explanation given above is compared with the "Genesis" explanation of how we humans came to exist, it is not surprising that the theory of evolution has caused so much controversy.

to explain evolution fully. This leads to a section on *Theistic Evolution* with a consideration of its benefits and drawbacks. Finally, I address the problem of cruelty in the evolutionary process and I also deal with the question, "Does theistic evolution introduce a *God of the gaps?*" I end with a closing discussion and some conclusions.

3.2 Darwinian Evolution

The famous biologist, Charles Darwin, was born in Shrewsbury in Shropshire in 1809. After studying medicine at Edinburgh University and biology at Cambridge University, he set out, in 1831, on a five-year scientific expedition on the survey ship HMS *Beagle*. The main purpose of this voyage for the *Beagle* was to carry out detailed hydrographic surveys around the southern part of South America. The *Beagle*'s captain, Robert FitzRoy, thought that it would be a good idea to have an expert geologist on board so he invited Darwin to be his travelling companion. Since he was keen to visit the tropics the young Charles Darwin duly accepted the invitation. His ideas on evolution began to take shape fully when the *Beagle* visited the Galapagos Islands in the east Pacific Ocean. Here he found that he could examine a unique and diverse range of animal life and find numerous examples to illustrate his theory that evolution occurred due to a process of natural selection. This theory proposed that the organisms best suited to their environment are more likely to reproduce and pass on the characteristics which helped the survival of their species. Gradually the species changes over time.

Natural selection can be considered in the following way:

1. Over long periods of time small changes randomly occur in species.

2. Some of these changes may give advantage for survival to the offspring in their living environment.

3. When this happens it looks as though nature had 'selected' those characteristics that are beneficial for survival of the species.

4. Those species that do not have characteristics beneficial for survival are more likely to become extinct.

This was a revolutionary view that provided, what appeared to be, a very different account of the development of life than the one that could be found in early chapters of the Genesis chapter in the Christian Bible.

When he returned to England Darwin committed himself to writing up his theory but it was not until 1859 that he published his famous book, *The Origin of Species by Means of Natural Selection (79)*. One of the main reasons for this delay was that Darwin was worried about the controversy his findings would have on the fundamental teachings of the Church in England. While, surprisingly, his work was initially accepted by the Church of England, he was right to be apprehensive and there has been lots of controversy over his theory up until the present day. However, in the scientific community, his theory is widely accepted and his invaluable research is widely praised.

3.3 Three Viewpoints on Human Development

Again, despite its success major controversy still surrounds the questions as to whether evolution by itself provides the best explanation of how humanity has developed. The significance of this explanation and how it relates to a belief in a creator God is also a subject of much dispute. Some supporters of *atheistic evolution* present it as a scientific alternative to God. However many theists, a substantial portion of them scientists, advocate *theistic evolution* and are ready to accept the theory of evolution which they see as God's method of creating the wide variety of life in the world. Other theists have different levels of willingness to accept evolution and advocate *Intelligent Design* while others, such as *Young Earth Creationists,* reject evolution completely.

Young Earth Creationism

Young Earth Creationists (YEC) claim that the earth was created some 10,000 years ago. They also believe that God made the universe and all original forms of life on Earth in six (24-hour) days. In other words they believe in a literal interpretation of the opening chapters of Genesis. The timescales they suggest for these processes are far too short to allow evolution to work so they reject the interpretation of the evidence given by evolutionary scientists. The prevailing attitude of Young Earth Creationists and the other groups who share similar views, is reflected in the following statement from Professor Henry Morris. "When science and the Bible differ, science has obviously miscalculated its data" *(80)*. Professor Morris is a civil engineer who, last century, founded and subsequently became President of the Institute for Creation Research. This Institute is a major contributor to the work of the Young Earth Creationists (YEC).

Commenting on views of those who reject evolution, the leader of the Human Genome Project, Francis Collins *(81)* has stated: -

"The shelves of many evangelicals are full of books that point out the flaws in evolution, discuss it only as a theory, and almost imply there's a conspiracy here to avoid the fact that evolution is actually flawed. All of those books, unfortunately, are based upon conclusions no reasonable biologist would now accept."

I also believe that YEC have got things badly wrong and it is the early chapters of Genesis which need careful and perceptive interpretation. To help provide this interpretation, for a start, I can recommend a book by John Lennox entitled *Seven*

Days that Divide the World (82). The scientific evidence contradicting the claims of the YEC, particularly the fossil records found in the world's rocks, is so overwhelming that I thought twice about even including the YEC claims in this review. However, the fact that in 2011 a Gallup survey held in the United States reported that over 30 per cent of its adult population interpret the Bible literally *(83)* persuaded me that YEC claims should be included. Nevertheless I feel compelled to stress that, as a scientist, I find that the views of the YEC unrealistic. I also consider that the strict adherence to these views by a number of Christian groups deters a substantial number of people from the Christian faith.

Intelligent Design

One of the earliest and most quoted proponents for Intelligent Design, (ID), was the 18th century English clergyman and philosopher William Paley. He was responsible for what has become known as the Watchmaker Analogy. The thinking behind this analogy is made clear in the following statement from his book *Natural Theology* **(84)**: - "In crossing a heath, suppose I pitched my foot against a stone, and were asked how the stone came to be there; I might possibly answer, that, for anything I knew to the contrary, it had lain there forever: nor would it perhaps be very easy to show the absurdity of this answer. But suppose I had found a watch upon the ground, and it should be enquired how the watch happened to be in this place; I should hardly think of the answer I had before given, that for anything I knew, the watch might always have been there (----) There must have existed, at some time, and some place or other, an artificer or artificers, who formed the watch for the purpose which we find it actually to answer; who comprehended its construction, and designed its use (....) Every indication of contrivance, every manifestation of design, which existed in this watch, exists in the works of nature; with the difference, on the side of nature, , of being greater or more, and in a degree which exceeds all computation."

Commenting on the amazing adaptations of plants and animals, Paley expanded this argument to claim that the complex structures of all living things required a designer. That designer was God.

In my experience most of those committed to Intelligent Design reject evolution and contend that living things are best explained by an intelligent cause and not an undirected process such as natural selection. They maintain that there are many complex biological phenomena that are too sophisticated to have evolved through a series of chance mutations. A particularly good example of one of these phenomena is the human genome which is built as instructed by the language of DNA. I will deal with this further in Chapter 4 where the information provided by DNA is shown to be like that given in an instruction book for how every living thing is constructed. In another example proponents of ID put forward the view that the complexities of even just one single living cell should be an indication that an intelligent designing mind is at work.

Initially I found the arguments put forward for ID very appealing. However, in recent years it has lost ground to the arguments based on Darwinian evolution, as far as human development is concerned. I believe that the main reason for this is that the theory of evolution is solidly supported by extensive and compelling scientific evidence such as that observed in Donald Prothero's excellent study, *Evolution – What the fossils say and why it matters (85)*. However, another reason emerges from the fact that Intelligent Design tends to get confused with the views of fundamentalists like the Young Earth Creationists as I discussed earlier. It should also be noted that some ID proponents accept that evolutionary forces[12] operate in the world but they are far from being the whole story. As I indicate later I found that I had considerable sympathy with this viewpoint.

Evolution

We have noted that the theory of evolution is grounded in Charles Darwin's *Theory of Natural Selection (79)* which states that modern species are products of an extensive process that began 3 billion years ago with simple single-celled organisms. The natural selection process provides the primary

---

[12] An evolutionary force can be defined as any factor that brings about changes in gene frequencies or chromosome frequencies in population and is thus capable of causing evolutionary change. Forces listed include, founder effects, genetic drift, mutation, migration and selection.

mechanism within this theory and is the result of genetic and environmental forces acting on organisms. Through the survival of the most adaptable species over very long periods of time, humans have developed and evolved to arrive at their present condition.

On the molecular scale evolution can also be explained starting with a few subatomic particles and some general Laws of Nature. First, obeying general laws, the subatomic particles assemble into stable atoms. These atoms assemble into long, complicated self-replicating molecules which form codes for assembling proteins into organic bodies. These bodies replicate. During replication mutations occur and, over a very long time, the bodies become increasingly well adapted to the environment so that complex organisms might live and reproduce.

The overwhelming support for evolution from scientists is not surprising when recent research is considered. Current scientific reviews of data, obtained from the fossil records, report on reliable and extensive evidence which give solid backing for the theory.[13]

---

[13] An outstanding work which helps to provide this support is a recent book, *Evolution – What the Fossils Say and Why It Matters* (**85**), written by the geologist Professor Donald R. Prothero, who also lectures in geobiology at the California Institute of Technology. Professor Prothero has provided a splendidly informative study covering finds which include some of the tremendous fossil discoveries of the past 20 years or so. Perhaps it has been caused in part by my training but I have always had a particular liking for data presented in visual form. For me, interpretation of one page of graphs can often prove much more enjoyable and immediately informative than the same information provided on several pages of mathematical

Two recent reports on "Views on evolution among the public and scientists" *(86)* have been published in the United States by the Pew Research Centre in Washington and the American Association for the Advancement of Science. In the first report it is recorded that:-

"In 2009, 97% of scientists and 61% of the public accepted evolution while 2% of scientists and 31% of the public rejected evolution. Among the scientists who accepted evolution, 87%, attributed it to natural processes and 8% to divine guidance: among members of the public who accepted evolution, 32% attributed it to natural processes and 22% to divine guidance."

In the second report, carried out in 2015, virtually the same overall result was obtained with 98% of scientists supporting evolution and 65% of the public supporting evolution.

Despite this strong support for evolution there are further areas of controversy. Some biologists consider that evolution can be split into microevolution and macroevolution. Roughly speaking, studies involving macroevolution focus on the change that occurs at or above the level of species while studies involving microevolution refer to smaller evolutionary changes. Evolutionary theorists such as Richard Dawkins, who consider that the evolutionary process should be treated as a

---

analysis. It is not surprising therefore that I found the illustrations in this book fascinating. The book is divided into two parts, Part 1 deals with *Evolution and the Fossil Record* while Part 2 is titled, *Evolution? The Fossils Say YES!* So it is clear where Donald Prothero's sympathies lie in the creationism/evolution debate.

continuous whole, have some reservations about this distinction. They contend that macroevolution results from microevolution processes operating over extremely long periods of time. While others, such as John Lennox *(87)* think that while the continuous process can explain the selection mechanisms which reasonably account for variations in finch beak lengths, it cannot account for the existence of finches or bacteria in the first place.

John Lennox then goes on to produce a rather telling statement from Paul Wesson, an eminent professor of astrophysics and theoretical physics from the University of Waterloo in Canada. The statement is as follows: -

"Large evolutionary innovations are not well understood. None have ever been observed, and we have no idea whether any may be in progress. There is no good fossil record of any."

This contrasts with our knowledge of microevolution where there is ample evidence of variations due to mutation and natural selection. Perhaps better evidence for macroevolution has been produced since this statement from Professor Wesson and geologists like Donald Prothero have given enthusiastic support for it, but here is little doubt that at present the claim that microevolution "flows" to macroevolution in a continuous process has not been proved.

While I think that this problem needs to be resolved, I believe that current science presents a compelling case for the importance of the contribution of evolution to the development of human beings. Although the contribution may be more limited than that claimed by some of the more adamant advocates of evolutionary theory.

3.4 Does belief in God prevent acceptance of the theory of evolution?

At present many involved in science contend that, since evolution is seen to be a "blind, undirected process", it cannot be believed to be the handiwork of a creator God. Eminent scientists such as chemistry professor Peter Atkins go further, expanding the argument to include the whole of science. This prompts them to make statements such as *(88)*, "It is not possible to be intellectually honest and believe in gods. And it is not possible to believe in gods and be a true scientist." Not surprisingly the contention, that you cannot be a true scientist and believe in God, has caused much argument and the ethologist Richard Dawkins is at the heart of the present-day controversy. While I do not believe that Professor Dawkins has ever directly claimed that scientific theory, or more specifically the theory of evolution, can disprove the existence of a creator God, there is no doubt that his strong and effective support for what we might call atheistic evolution, has led to the commonly held view that this is the case. The situation is not helped by the fact that a substantial number of people have also formed atheist views in reaction to the unconvincing attacks on evolution put forward by Christian

Fundamentalists, such as the Young Earth Creationists.

While I believe in a creator God, when scientific arguments for and against evolution are compared I am persuaded by the case put forward by evolutionists. Unfortunately this success of evolutionary theory has brought with it the confusion that science can disprove the existence of a creator God. As reported in Chapter 1, it was this claim which led to one of my own doubts about the existence of a creator God. I shall now examine the validity of this doubt.

Even a brief look at the amazing advances in science, such as those mentioned in the early chapters of this book, help to show why science has had such a powerful influence on our picture of reality. However, we must be careful when using science to explain things. There are limits to what science can explain and this must be realised when it is used to support argument since, as shown in the following examples, there are cases when the explanation it provides are not be very helpful to the questioner. This was first illustrated to me during a lecture, concerned with the strength of materials, which I attended at the University of Glasgow many years ago. During the proceedings, the lecturer, who I remember as being rather entertaining despite his dull subject, related a cautionary tale concerning one of our aged professors. We shall call him Professor McDuff, a professor who was devoted to his subject of materials science.

McDuff lived by the sea and on a sunny day he

liked nothing better than taking a bus trip along the Ayrshire coast road from Ayr to Girvan. He was particularly happy if the bus was double-decked and there were no children on board because then he could take his seat upstairs at the front and get a great view of all that was happening outside the bus. On one such trip he was sitting happily looking out of the window when he noted a large wheel rolling along beside the bus. He quickly realised that the wheel had broken free from the back axle of the bus and, displaying an agility which defied his years, he quickly ran downstairs and jumped from the bus before it came to a stuttering and grinding halt. He wasted no time in reaching the broken axle and began a thorough examination. Meanwhile the driver pulled himself, dazed and shaken, from the cab.

"Why did that happen?" he asked.

"I'm not sure," replied Professor McDuff as he peered at the edge of the broken shaft, "but I think it's a fatigue fracture."

"Oh!" replied the driver. "I thought that the wheel might have come off!"

On a more serious note, when it comes to 'why' questions, science has problems. John Lennox [14] states, "There are certain questions that science is not geared to answer, particularly 'why' questions that have to do with purpose as distinct from function."

---

[14] The source of the information presented here can be found in *God's Undertaker* by John Lennox *(89)* in a section where he assesses carefully and perceptively the limits of scientific explanation. As with most of Lennox's explanations, I have found the examples given in this section of great help in my study.

To support this statement he presents a simple illustration which makes the use of a newly baked cake, his Aunt Matilda and a number of expert scientists. He asks us to imagine that his Aunt Matilda has baked a beautiful cake which he takes along for analysis by a group of the world's tops scientists. On arrival he asks them to provide an explanation of the cake and they go to work. When their analysis is complete the nutrition scientists can report on the nutritional effect of the cake and the number of calories it contains; the biochemists can tell us about the stricture of the proteins, fats, etc. in the cake; the chemists will have examined the elements involved in their bonding; the physicists will have completed an analysis of the cake in terms of fundamental particles; and the mathematicians will have derived a set of equations to describe the behaviour of these particles.

When the analysis is complete, each expert, in terms of his or her scientific discipline, will have provided an exhaustive description of the cake but can we claim that we have a full explanation? We now have a description of *how* its various constituent elements relate to each other, but we still do not know *why* the cake was made and the experts cannot provide an answer. Their disciplines can cope with questions about the nature and structure of the cake, that is answering the *how* questions. The disciplines cannot answer *why* questions, connected with the purpose for which the cake was made. The only way to get an answer is if Aunt Matilda reveals it to us and without her cooperation no amount of scientific analysis will enlighten us.

By thinking about this example we are helped to see that accepting evolution as a sound explanation of *how* we got here does not prevent belief in a creator God which answers *why* we are here. So in answer to the first question which I asked at the start of this chapter, belief in God does not prevent acceptance of evolution. Importantly, we can also note that science by itself cannot be used to explain evolution fully. It cannot explain why the evolution process was initiated.

3.5 Does evolution need God?

I have noticed that, recently, scientists who do not believe in God seem to be moving away from the assertion "God does not exist" to the assertion "There is no need for God". For instance, in his recent book *The Grand Design (90)* Stephen Hawking asserts that Darwin "explained how the apparently miraculous design for living forms could appear without the intervention of a supreme being". In his book *God's Undertaker. Has science buried God? (5)* John Lennox has responded to this sort of assertion. In a section titled "God – an unnecessary hypothesis?" he shows the flaws in the reasoning employed by scientists to claim that God is unnecessary. To accomplish this he uses an illustration which involves a Ford motor car[15].

---

[15] This rewarding and superbly insightful illustration is presented in *God's Undertaker (91)* where the author, John Lennox, asks, "God – an unnecessary hypothesis?"

Aided by my background in science and engineering, I find Lennox's illustration both insightful and rewarding. He begins by asking us to consider the reaction of someone from a remote part of the world on seeing a motor car for the first time. If this person knew nothing about modern engineering he might imagine that there was a god inside the engine, making it run. He might also imagine that when the engine was running well it was because Mr. Ford liked him and when it refused to go it was because Mr. Ford did not like him. However, if he were to study engineering and dismantle the engine, he would discover that there was no Mr. Ford inside it. He would also see that he did not need to introduce Mr. Ford as an explanation for its working. His newly acquired knowledge of the principles of operation of the internal combustion engine would explain things. However, if he decided that his knowledge of how the engine works made it impossible for him to believe in the existence of a Mr. Ford who designed the engine in the first place, this would be patently false. Had there never been a Mr. Ford to design the mechanisms, none would exist for him to understand. Lennox notes that in philosophical terms he would be committing a category mistake.

Clearly, it is committing the same category mistake to assert that our understanding of the scientific principles of evolution make it unnecessary for us to believe in a creator God. Again to quote John Lennox, "we should not confuse the mechanisms by which the universe works either with its cause or its upholder."

It is also worthwhile noting that several leading scientists, including Dawkins among them, seem to consider that it is permissible to use God in an explanation in direct competition with using science in the explanation. For example, in a recent discussion reported in SchansBlog *(92)*, Dawkins is quoted as stating: "Evolution is the creator of life and life is arguably the most surprising and beautiful production that the laws of physics have ever generated." I would contend that here Dawkins is confusing law on the one hand with agency on the other. Laws by themselves cannot produce life. For that we need an agency and I believe that that agency is a creator God. Richard Dawkins does not agree with this belief and in his book *The Blind Watchmaker (93)* he contends that, "Theistic Evolution is a superfluous attempt to smuggle God in by the back door." However, I found further agreement for my position from Keith Ward, formerly Regius Professor of Divinity at the University of Oxford. Keith Ward considers that supernatural intervention in the evolution process is more than likely and in his book, *Why There Almost Certainly is a God (94)*, he replies to this contention from Dawkins making the following statement on the evolution process:-

"We do not need an intelligent creator. Blind and gradual selection will do the trick. Well, it may and I do not deny it. But the chances of doing so seem to be astronomically small. Unless that is that the laws governing the sorts of mutation that occur have been carefully worked out beforehand."

He goes on to contend:-

"It is not true that the postulate of an intelligent creator is superfluous. For such a creator would raise the possibility that the process would result in intelligent life by an enormous amount. In fact it would make it virtually certain, as opposed to being one possibility among others."

I welcomed this support for my claims for theistic evolution. For me, the origin of life, the start of the evolution process, provides the biggest mystery. Even Richard Dawkins *(95)* in his book *The God Delusion* refers to the origin of life as "that initial stroke of luck." I do not believe it was luck but the act of a creator God. I would like to stress again that the theory of evolution cannot explain why the process of evolution started. Therefore it cannot be used to explain evolution fully. It cannot even attempt to say why we are here. In my view theistic evolution provides the best explanation currently available.

3.6 Theistic Evolution

What is theistic evolution?

My views on theistic evolution are shared by Francis Collins *(96)* who writes, "....I believe that God had a plan to create creatures with whom he could have fellowship, in whom he could inspire (the) moral law, in whom he could infuse the soul, and who he would give free will as a gift to make decisions about our own behaviour, a gift which we oftentimes utilise to do the wrong thing. I believe God used the mechanism of evolution to achieve that goal."

In his book *The Language of God (97)* Collins puts forward an enthusiastic case for theistic evolution, maintaining that it is the dominant position of serious biologists who are also serious believers. He points out that:-

"There are many subtle variants of theistic evolution, but a typical version rests upon the following premises; -

1) The universe came into being out of nothingness, approximately 14 billion years ago.

2) Despite massive improbabilities, the properties of the universe appear to have been precisely tuned for life.

3) While the precise mechanism of the origin of life on earth remains unknown, once life arose the process of evolution and natural selection permitted the development of biological diversity and complexity over very long periods of time.

4) Once evolution got underway, no special supernatural intervention was required.

Humans are part of this process, sharing a common ancestor with the great apes.

But humans are also unique in ways that defy evolutionary explanation and point to our spiritual nature. This includes the existence of the moral law (the knowledge of right and wrong) and the search for God that characterises all human cultures throughout history."

I agree when Francis Collins contends that if we accept these six premises "then an entirely plausible, intellectually satisfying and logical synthesis emerges." However, before committing myself fully there were two main questions which needed to be addressed. First, why is evolution such a cruel process? Second, is theistic evolution simply introducing a "God of the gaps"?

Why can evolution be such a cruel process?

Before completely accepting theistic evolution I felt that I had to deal with the fact that evolution can be a cruel process. I was vexed by the question of how such a process could be claimed to be the work of a caring God. There is no doubt that the process of evolution can prove to be very cruel indeed. As a Christian, I find it very difficult to find an answer for this cruelty and later in this book I deal with moral issues at some length. In the Appendix, after giving having my reasons for belief in a creator God I consider the nature of this God and I look further at the problem of suffering with some help from the James Gregory lecture given by John Polkingthorne *(98)* at St Andrews University in 2008, *Does God Interact with his Suffering World?*

As stated in Chapter 1, at this stage in my study I am using hard scientific facts to help us think about the possibility that God exists. Scientific explanation alone cannot distinguish between good and evil and moral issues are not relevant at this point of development in my argument. At this stage I am not trying to decide whether God is good, bad or indifferent. This will be done later. It is not helpful to try to answer two questions at once. Here my first question is, "Does a creator God exist?" The fact that God, if he exists, can allow cruelty on this Earth does not present a valid reason for saying that there is no God. When I have an answer to my first question, my second question, which concerns the moral nature of God, can be addressed.

Does Theistic Evolution introduce a "God of the Gaps"?

Often when discussing theistic evolution the issue of "God of the Gaps" is raised. This is the idea that the introduction of God is caused by intellectual laziness; we cannot fully explain evolution scientifically and we introduce God to cover our ignorance. To answer this criticism I return to John Lennox and his astute illustration, employing Mr. Ford *(91)*. Here Lennox points out the importance of the fact that Mr. Ford Is not to be found in the gaps of our knowledge about combustion engines. In fact he is not to be found in any reason-giving explanations that concern mechanisms. He is not a mechanism: he is the agent responsible for the existence of the mechanism and it bears all the marks of his handiwork, and that includes the bits we do understand and the bits we don't.

The following eloquently expressed explanation

from the philosopher Richard Swinburne in his book, *Is there a God? (99)*, says it all: -

"I am not postulating a "God of the gaps", a god merely to explain things that science has not yet explained. I am postulating a God to explain why science explains; I do not deny that science explains, but I postulate God to explain why science explains."

3.7 Discussion

The information gained from my study has shown that we cannot use evolution to explain why we are here. In Section 3.3, I pointed out that the science of evolution is not geared to answer 'why' questions: questions that have to do with purpose as distinct from function. While the theory of evolution can explain 'how' humans have developed into our current state it cannot explain 'why'. I also discovered that acceptance of the science of evolution does not prevent belief in the existence of a creator God and the theory of atheistic evolution cannot fully explain the evolutionary process.

To help decide on the best explanation of the development of human beings I first looked at the choices of explanation available and I elected to consider the three viewpoints that I estimated had received the most attention in the literature, *Young Earth Creationism, Intelligent Design* and *Evolution.*

Young Earth Creationists contend that evolution did not take place and a literal interpretation of the first pages of the book of Genesis in the Old Testament of the Christian Bible gives a reliable description of how human beings of the contemporary worlds were created. In the United States some 30% of the adult population claim to believe in literal interpretation of the Bible. However, as a scientist, I find this unacceptable. Also, as a Christian, I find it positively damaging.

As a theist I believe in Intelligent Design (ID), but as far as the explanation of evolution is concerned I do not put myself in the ID camp. My position as a Theistic Evolutionist can be neatly summarised by Ken Miller, a professor of biology at Brown University in Rhode Island, who says that Theistic Evolutionists believe that natural processes provide necessary and sufficient support to bring about the origin of all living things while adherents of Intelligent Design do not believe that natural processes are sufficient *(100)*. My belief in a creator God has not been undermined by an awareness of the cruelty of evolution or the possible suggestion that I am introducing a "God of the gaps". In conclusion, I contend that theistic evolution gives the best explanation of how we got here.

3.8 Key Points

In sum the main points considered in this chapter are as follows: -

- Science shows that the views, on evolution, of Young Earth Creationists are not valid.

- The theory of evolution gives the best explanation of how we got here.

- The science of evolution cannot fully explain human development.

- Acceptance of the theory of evolution does not prevent belief in the existence of a creator God.

- I believe Theistic Evolution provides the best explanation of the evolution process.

CHAPTER 4

Science and Reality

The universe does not have just a single existence or history, but rather every possible version of the universe exists simultaneously in what is called a quantum superposition.

*- **Stephen Hawking** (Professor of Mathematics, Cambridge University)*

*Let us recognize these (many-universe) speculations for what they are. They are **not** physics, but in the strictest sense, metaphysics.*

*- **John Polkinghorne** (Fellow of Queens College Cambridge University)*

I think I can safely say that no one understands quantum mechanics.

*- **Richard Feynman** (Nobel Prize physicist)*

4.1 Introduction

Science has an enormous influence on our lives in the contemporary world. Virtually everything we do is affected by spectacular advances in science, particularly in physics, computing and information technology, medicine and biotechnology. This amazing success has given science a dominance in our thinking. There are many people today who seem to believe that science can provide "all the answers". If not immediately, then at some point in the future.

The reaction of some physicists to the recent discovery of the subatomic particle named the *Higgs boson* helps to illustrate this confidence. This particle has become popularly known as the "God particle" much to the disapproval of most physicists. However, while he dislikes this name, the theoretical physicist Lawrence Krauss (Professor of the School of Earth and Space Exploration at Arizona State University) has written *(102)*: - "Humans with their remarkable tools and their remarkable brains, may have just taken a giant step towards replacing metaphysical speculation with empirically verifiable knowledge. The Higgs particle is now arguably more relevant than God."

While I would take issue with Krauss' statement (we shall return to this in the following sections), it does provide a clear example of the high level of confidence that some people have in science[16].

However, during my study on materialism I became increasingly concerned about the reliability of using natural science to help explain reality. Scientists did not seem able to reach agreement, or even give a satisfactory answer, on some of the more important problems now facing them. I decided to consider this further. I uncovered much information of direct relevance to my study and I shall present some of my findings in this chapter.

First I take a look back at the last century and note how some key scientific advances have helped to provide a new picture of physical reality. I report that advances in biology over this period have enriched our knowledge of what we really are and the physical sciences have produced an amazingly different picture of the physical reality of the universe. After reflecting on these advances, I argue that science has still not given us a completely reliable explanation of reality. I consider some of the entrenched beliefs held by many contemporary scientists and argue that if we are to find the truth their present mind-sets must change. To find acceptable answers to several important problems, I argue that scientists need to pay more attention to the works of relevant philosophers. I then consider some intriguing problems being faced by current

---

[16] To avoid any ambiguity it should be noted that the definition of science used in this text is as follows:-
"The intellectual and practical activity encompassing the systematic study of the structure and behaviour of the physical and natural world through observation and experiment." **(101)**

scientists in their search for the truth. This leads me to an assessment of the current limitations of science. Finally, I close with a discussion which reaches some firm conclusions on the ability of science to help explain what is real.

4. 2. Advances in science – emergence of a new Reality

In this section, I look back to the middle of the 19th century to consider some of the major advances in science back then – advances that are particularly relevant to the present study. I deal with these advances in two sections. The first section deals with advances in biology and the second focusses on developments in the physical sciences. As we shall see, science has undergone immense and unforeseen changes during the past two centuries and, as we continue to advance, it seems probable that further major surprises await us.

Biology

At one point in my research career I became interested in heat transfer in the bioreactor vessels used in the large-scale production of biochemicals and I was soon astounded by the momentous changes taking place in the field of biotechnology. In the field of biology, there is little doubt in my mind that the two most important advances have concerned developments with the theory of evolution, first proposed by Charles Darwin in 1859, and the deciphering of the structure of deoxyribonucleic acid (DNA) by James Watson and Francis Crick in 1953[17].

In Chapter 3 we have already considered advances concerning evolution but, so far, I have said little about DNA which can be described as, "A self-replicating material which is present in nearly all living organisms as the main constituent of chromosomes. It is the carrier of genetic information."*(104)* Most of us will be aware of the familiar double helix coiled structure of the DNA molecule[18]. DNA is of major importance to biotechnologists because it contains the blueprints for the manufacture of proteins. Proteins are the basic biochemicals of living organisms and determine their development and all the processes of life. The structure of protein is based on units called amino acids. I believe that the relationship between DNA, amino acids and proteins can best be described by using a musical analogy which involves a piano, a pianist and a sheet of music placed at the piano keyboard. The sheet of music represents DNA and

---

[17] In 1962, Francis Crick and Harry Watson from the Cavendish Laboratory Cambridge, along with Maurice Wilkins from Kings College, London, won the Nobel Prize in Medicine for their discovery of the structure of DNA **(103).** This is universally accepted as one of the most significant scientific discoveries of the 20th century.

[18] The gene can be defined as "a hereditary unit consisting of a sequence of DNA that occupies a specific location on a chromosome and determines a particular characteristic in an organism".

the amino acids are the notes that the pianist can select. When the right notes are played in the right order shown on the music sheet the result is a melody – the protein.

The fact that genes are made up of DNA which can be isolated, copied and manipulated, has led to the astounding advances of modern biotechnology particularly in the fields of medicine, agriculture and forensics. In the field of medicine, gene modifications are used in the production of therapeutic human proteins, such as human insulin, and modern biotechnologies often involve manipulating vaccines so that they are more effective or can be delivered by different routes. Gene therapy technologies are now being developed to treat such diseases as cancer, Parkinson's disease and cystic fibrosis. In agriculture plants and animals can be improved by genetic modification with beneficial traits being identified by DNA profiling. The use of DNA In forensic analysis has been well publicised and the identification of DNA samples at a crime scene, or for the determination of parentage, have proved to be of great assistance. Recently I took some interest in a bioremediation project where organisms were used to clean up polluted soil and was very impressed by the effectiveness of the organisms in clearing really nasty waste.

The cloning of Dolly the sheep in the mid-1990s by Ian Wilmut and Keith Campbell[19] at the

---

[19] Dolly, a Finn Dorset sheep, was born on July 5th, 1996, at the Roslin Institute in Edinburgh. Dolly was world's first mammal to be cloned from an adult cell and her birth is considered to be one of the most significant scientific breakthroughs ever. The team that created

University of Edinburgh, certainly proved to be a major event in the field of biotechnology when Dolly became the first mammal to be cloned from an adult cell. However, for the purposes of our study on the nature of reality I would suggest that most significant recent advance in life sciences was the completion, in the year 2000, of the ground-breaking Human Genome Project (HGP). This gave us an amazing code carrying with it all of the instructions for building a human being.

This project was an international collaborative research programme involving some twenty research laboratories from the United States of America, the United Kingdom, Japan, France, Germany and China. The main aims of this project concerned the complete mapping and understanding of the human genome. A clear description of the project can be found in reference *(106)*. The start of this research can be traced back to the work of an undergraduate researcher, Alfred Sturtevant, working at the Morgan Laboratory in Kansas in the USA in the year 1911 and the end can be considered to be the completion of the Human Genome Project, with publication of the first draft of the human genome in the journal *Nature* in 2001 *(107)*.

Dr Francis Collins was director of the final phase of the HGP. During this project he led a team of international scientists, in a programme of research which lasted more than ten years. Describing a summer morning which marked completion of the

her was led by Scotsman Ian Wilmut from Edinburgh University **(105)**.

project, Dr Collins wrote the following in the introduction to his book *The Language of God (7)* :-

"The human genome consists of all the DNA of our species, the hereditary code of life. This newly revealed text is 3 billion letters long, and written in a strange cryptographic four letter code. Such is the amazing complexity of the information carried within each cell of the human body that a live reading of that code at a rate of one letter per second would take thirty-one years, even if the reading continued day and night. Printing these letters out in regular font size on normal bond paper and binding them all together would result in a tower the height of the Washington Monument. For the first time on that summer morning (when the work was completed) this amazing script, carrying with it all of the instructions for building a human being, was available to the world."[20]

I deal further with the significance of this momentous project in Section 5.3 of the next chapter.

Before completing this section, I thought that it would be useful to take a brief look into the future to see what it might hold for evolution. Scientific prediction of the future development of the evolutionary process is fraught with difficulties. It is generally agreed that evolution is chaotic and unpredictable. For us the direction of human evolution is of enormous concern particularly when

---

[20] This quote can be found in the first two pages of reference *(7)* where Collins introduces an absorbing and informative study which is divided into three parts – Part One: The Chasm between Science and Faith, Part Two: The Great questions of Human Existence, and Part Three: Faith in Science and in God.

we consider the potential influence of our advancing knowledge of genetics. Some attempts have been made to speculate on the possible development of evolution in order to warn us of potential dangers. A good example can be found in a recent paper, "The Future of Human Evolution" by Professor Nick Bostrom **(108)**, of the Future of Humanity Institute in the Faculty of Philosophy and Oxford Martin School at the University of Oxford. Professor Bostrom introduces his paper as follows:-

"Evolutionary development is sometimes thought of as exhibiting an inexorable trend towards higher, more complex, and normatively worthwhile forms of life. This paper explores some dystopian scenarios where freewheeling evolutionary developments, while continuing to produce complex and intelligent forms of organisation, lead to the elimination of all forms of being that we care about."

Scientific speculation on future human evolution has led to a number of different claims. Some scientists, such as Steve Jones, a genetics professor at University College, London, contend that we have stopped evolving *(109)*, while others such as Geoffrey Miller, an Associate Professor at the University of New Mexico, believe that Darwinian evolution in human beings is speeding up *(110)*. Adding to the available options, I recently had the pleasure of listening to molecular biologist Denis Alexander at a conference held in the University of St Andrews. In his paper, entitled "Creation, Providence and Evolution" *(111)* Alexander presented some intriguing evidence that human evolution might not be as unpredictable as commonly believed. However, it soon became clear to me that any speculation about the future of human evolution must involve consideration of advances in the field of genetics, and any predictions of future advances must involve consideration of the combined effects of evolution and genetics.

In recent years numerous advances have taken place in genetics concerned with medical research and investigations involving stem cells and cloning have a high profile. Most researchers involved in this work are highly optimistic about the successful use of genetics to cure a whole range of medical ailments. At present, exciting research using stem cells is advancing the treatment of diseases such as Parkinson's, chronic heart disease, leukaemia and other illnesses. Looking further into the future, further major advances can be anticipated. Cloning and bioengineering of body parts leading to transplants also show tremendous promise.

These advances prompt extremely difficult ethical decisions for scientists, particularly in processes of unnatural selection, which could lead to the success of *transhumanism* where humans take charge of evolution and transcend their biological limitations by using technology[21]. Achieving success with the scientific aspects of transhumanism will involve major advances in genetic science but the results of current research involving stem cells and cloning show that our ability to make changes

---

[21] The objective of the intellectual and international movement, *Transhumanism,* is to transform the human condition by developing and creating available technologies to enhance human intellectual, physical and psychological capacities.

in the human condition is making startling progress

So, summarising this section, modern advances in biology have helped to improve our knowledge of reality. Progress on evolution has revolutionised our thinking on human development and amazingly, now we have the complete genetic information needed for the creation of a human being. However, the increasing developments in transhumanism are worrying and give us enormous ethical concerns and, as discussed later, science by itself is of limited use when dealing with matters of morality.

Physical Sciences

Towards the end of the 19th century William Thomson, who became Lord Kelvin, was Professor of Natural Philosophy at my old university, the University of Glasgow. When I studied there in the 1960s Lord Kelvin, like the even more famous James Watt, was a bit of a hero at the University's Mechanical Engineering Research Annexe, where an ambitious programme on the measurement of the thermodynamic and transport properties of steam was the main focus of our research work. However, even brilliant researchers like Lord Kelvin could get things wrong and he got things very wrong when he stated: - "There is nothing new to be discovered in physics now. All that remains is more and more precise measurement" **(112)**. Unfortunately for him, ever since he made that statement tremendous change has taken place.

This can be seen in the information provided in

earlier chapters. *Relativity, Quantum Physics and Chaos Theory* have given us a very different reality from the picture given by the Newtonian physics employed by Lord Kelvin.

Modern physics is divided into two parts, and each part represents a radical departure from the physics of the early 20th century. In one part, on the large scale, when objects move at great velocity, or are in the presence of strong gravitational forces, *relativity* dominates. It has been shown that the speed of light is constant and that space and time, should be considered together and in relation to each other. The significance of Albert Einstein's discovery in giving us a clearer picture of physical reality, particularly in the field of cosmology, has been overwhelming.

In the second part, by looking inwards towards the world of subatomic particles, we have seen in Chapter 2 that the main advances relevant to this study were *quantum physics, string theory* and *chaos theory*. These theories have introduced tremendous changes in our thinking.

We have noted that at present scientists cannot fully explain quantum physics. The truth of the situation is expressed by the often used quote from the exceptional Nobel Prize winning physicist Richard Feynman[22], "I think I can safely say that no one

---

[22] In my experience, **Richard Feynman** is now being increasingly revered as one of the "giants" of theoretical physics of last century. He held a number of highly prestigious awards, among them the 1965 Nobel Prize in Physics and he was elected as a foreign member of the Royal Society in London. He was a multifaceted individual with scientist, teacher, raconteur and musician listed among his abilities.

understands quantum mechanics." However, despite the lack of agreement, we are slowly realising a physical reality vastly different from what we considered to be real at the start of the 20th century.

The importance of string theory was emphasised in Chapter 2. If verified string theory will greatly alter our understanding of the nature of ultimate reality. It predicts that the fundamental particles of matter (quarks, electrons etc.) are composed of vibrating filaments of energy and contends a universe where all matter is generated by incredibly tiny loops of energy vibrating in many dimensions. These strings form the fundamental building blocks of matter. While recently this theory has run into difficulties it helps to indicate that advances in physical science are tending to give a much less "solid" picture of the matter in the universe.

I found further support for this claim when, on a recent holiday, I went browsing through a superb little bookshop, in the lovely little town of Aberfeldy here in Scotland. I was delighted to come across a new publication entitled *The Science Delusion (29)*. The author of this book is Rupert Sheldrake who started his career as a researcher in the field of biochemistry[23]. He became disenchanted

---

The often used quote given here **(113)** was made during the Messenger Lectures at Cornell University in 1964.

[23] **Rupert Sheldrake** majored in biochemistry during his undergraduate days at Cambridge University. Following this he spent some time at Harvard where he studied the history of the philosophy of science. He then returned to Cambridge to complete his PhD project which involved research on the development of plants. During the 1970s he was the principal plant physiologist at the International

by the strictly materialist approach of many scientists and now writes controversial but engrossing books on science. In *The Science Delusion* he has produced a highly informative and ground-breaking book which considers some of the present limitations of science and dismisses the mechanistic theory of the universe. In a chapter of his book entitled *Is Nature Mechanical?* Sheldrake supports and, I believe advances, the work of Davies and Gribbin. He concludes that living organisms provide better metaphors for organised systems at all levels of complexity and "in the light of the Big Bang theory, the entire universe is more like a growing, developing organism than a machine slowly running out of steam."

Consideration of the universe as "a growing developing organism" presents a very exciting and game changing concept. If our picture of reality can change so much in such a short period of time what other changes does the future hold? I believe that further major changes will come as science develops.

4.3 A More Reliable Explanation of Reality

I have already commented on the distrust that some scientists seem to have of philosophers. For instance

---

Crops Research Institute for the Semi-Arid Tropics. In recent years he has become increasingly interested in parapsychology.

Stephen Hawking gets no further than the first page of text in his book, *The Grand Design (23)* when he delivers a scathing criticism. Starting in the first paragraph he states:-

"How can we understand the world in which we find ourselves? How does the universe behave? What is the nature of reality? Where did all this come from? Did the universe need a creator? Most of us do not spend most of our time worrying about these questions, but almost all of us worry about them some of the time.

Traditionally these are questions for philosophy, but philosophy is dead."

I certainly do not go along with this distrust. In order to provide a more reliable explanation of reality, I contend that scientists need to give more respect and attention to non-scientific resources that can affect their work, such as philosophy. There are several intriguing problems that arise when thinking about the nature of reality, in which a satisfactory solution requires us to draw on philosophy. I shall deal with some of them here. First I deal with "some dogmas of modern science", pointing out that some scientists, with their entrenched attitudes, have adopted philosophical materialism. I then assess the progress being made towards the development of the elusive "Theory of everything". Next, I point out that we have not yet produced a satisfactory explanation of quantum mechanics without employing philosophy. Finally, I go on to mention a number of intriguing problems that are in need of a solution to help science contribute to a more reliable explanation of reality.

Some dogmas of modern science

In Chapter 2, I claimed that the philosophy of materialism is not valid but this point seems to have gone unnoticed by many scientists. I was delighted to find that Rupert Sheldrake strongly supports my views and has, for some time, been concerned with the overconfident and entrenched attitude of many of today's scientists. He argues: -

"Contemporary science is based on the claim that all reality is material or physical. There is no reality but material reality. Consciousness is a by-product of the physical activity of the brain. Matter is unconscious. Evolution is purposeless. God exists only as an idea in human minds, and hence in human heads.

These beliefs are powerful, not because most scientists think about them critically but because they don't. The facts of science are real enough; so are the techniques that the scientists use, and the technologies based on them. But the belief system that governs conventional scientific thinking is an act of faith, grounded in nineteenth-century ideology."

In, *The Science Delusion,* Sheldrake maintains that there are a number of 'dogmas' which constrict the progress of science and in a section entitled "The scientific creed" he lists, and questions, ten core beliefs which scientists take for granted. He argues that together these beliefs make up the philosophy or ideology of materialism centred on the assumption that everything is essentially material, even physical minds. He contends that the sciences will be regenerated when they are liberated from these dogmas. He then lists ten "beliefs". I now present four which are particularly relevant my own assertions: -

- "Everything is essentially mechanical. Dogs, for example, are complex mechanisms, rather than living organisms with goals of their own. Even people are machines, termed "lumbering robots", in Richard Dawkins' vivid phrase, with brains that are like genetically programmed computers."

- "All matter is unconscious. It has no inner life or subjective view. Even human consciousness is an illusion produced by the material activity of brains."

- "The Laws of nature are fixed. They are the same today as they were at the beginning, and will stay the same forever."

- "Nature is purposeless, and evolution has no goal or direction."

He then goes on to support his views with strong evidence and penetrating argument. While I do not fully accept Sheldrake's views on all of these 'beliefs', I agree with most of them and I was pleased to see that his objections to several of the listed 'beliefs' are supported by much of the information I have already presented in this book: particularly in Chapter 2, where I have reported that recent developments in the physical sciences have stressed the falseness of the belief that "everything is essentially mechanical".

At this stage I do not intend to go into any detailed analysis of Richard Sheldrake's assertions. I simply wish to bring his work to the attention of the reader and to applaud Sheldrake's efforts, which for me, illustrate how important it is to keep questioning the basic assumptions of a science which has been massively influenced by acceptance of the philosophy of materialism.

No 'theory of everything'

Early in my study, I came across a major problem for all of the physical sciences. Put simply it seems that the theoretical laws which apply at the large scale do not agree with the laws which govern the small scale. There is no single theory which unites the fundamental laws of physics. This seems to have come about because researchers in cosmology developed theories for their large scale projects and particle physicists did the same for their small scale research. It seems to have taken some time for them to get together to compare results.

To date no "ultimate theory of everything"[24] has been produced. Theoretical physicists have not yet

---

[24] *The theory of everything can be defined as a hypothetical single, all encompassing, coherent theoretical framework of physics that fully explains and links together all physical aspects of the universe. It is the "Holy Grail" of theoretical physics. The distinguished theoretical physicist Steven Weinberg describes it as the physicist's "dream of the final theory" **(116). 6).**

produced what our American cousins would call the "Real Deal".

In the recently published book, *The Grand Design* **(23)**, Stephen Hawking, assisted by Leonard Mlodinow, has put forward *M-theory* as a candidate for the ultimate theory of everything. In the first chapter of the book they explain:-"M-theory is not a theory in the usual sense. It is a whole family of different theories each of which is a good description of observations only in some range of physical situations. It is a bit like a map".

Perhaps I found the method, proposed by Hawking and Mlodinow, particularly appealing because it utilises the type of approach an engineer would use. The approach is called model-dependent realism. It involves collecting enough reliable observational data and then forming a mathematical model which gives a good description over a range of physical situations. When the physical range is extended, and the model ceases to be a good fit, additional relevant observational data are then collected for the extended range of situations to form another model. This second model is designed to have a region of overlap with the first. The process can then be repeated with a third region and so on. In this way a "map" covering a whole range of physical situations can be produced. Using this method M-theory can deal with ranges of physics such as those covered by quantum physics up to ranges best dealt with by cosmology.

In a book, titled, *God and Stephen Hawking – whose design is it anyway?* **(115)**, John Lennox has recently presented a concise and erudite criticism of some of the ideas expressed in *The Grand Design*. It is a clever little book which deals mostly with a criticism of the atheist claims made by Hawking and Mlodinow. However, while I applaud Lennox's findings, they do not challenge the validity of the science presented by Hawking and Mlodinow. While Hawking's book certainly appears to me to be a tour de force, it has to be conceded that M-theory is not the single theory which unites the fundamental laws of physics.

No satisfactory explanation of quantum physics

Providing a satisfactory explanation of the quantum world also presents a truly major stumbling block for scientific reliability. I was very surprised to discover that scientists on their own cannot agree on an interpretation for quantum mechanics. Ever since Niels Bohr proposed the Copenhagen Interpretation **(59)** many concerned with quantum physics have felt forced to turn to philosophy to find answers. I believe that it is also true to say that none of the existing interpretations proposed is considered to be totally satisfactory and at present we are left with making a choice between interpretations which predict subjective reality (as proposed by Neils Bhor), to suggestions where we have to believe each of us lives in an infinite number of universes (as proposed by the multiverse theory). Science will advance and we may find scientific solutions to present problems. However, as far as today's scientists are concerned, their answer to the question,

"What is ultimate reality?" must be, "We don't know." Here I believe that it is appropriate to note that earlier in this study I wrote, "Theists like me believe in a reality they do not understand." This statement about me can now be said to apply to quantum physicists who have had to turn to philosophy to try and fully explain their claims.

More intriguing problems

In the physical sciences there are several examples of other problems where our current scientific knowledge has yet to find a solution which would help expand the limits of our knowledge of reality. Here I would like to mention three which I think are particularly relevant to this study. Looking inwards towards the world of subatomic particles I suspect that at the very small scale, String Theory could produce results that will substantially change our views on reality. On looking outwards to the world of cosmology, where things also seem to be coming increasingly complex, I have become particularly intrigued by black holes, dark matter and especially dark energy. Clearly, science will need to make further advances before these phenomena can be fully explained and I consider that it is likely that these advances will lead to further changes in our concepts of reality.

A black hole is a region in space from which nothing can escape. It results from the deformation in space-time caused by an extremely compact mass. Infinite force of gravity exists at the centre of a black hole and as a result nothing, not even light, can escape. Around the circumference of the black hole we can place a two dimensional surface which marks the point of no return. This surface is called an event horizon.

Most of the predicted properties of black holes seem incredible. At the centre of the black hole is a *singularity* where, due to the infinite force of gravity, matter is crushed to infinite density. The laws of physics break down. Space and time are broken apart and cause and effect cannot be unravelled. At present there is no satisfactory scientific theory to explain reality "beyond" a singularity.[25]

I was particularly taken by one hypothesis on the *Holographic Principle,*[26] put forward by, among others, one of the world's leading scientists, Leonard Susskind, Professor of Theoretical Physics at Stanford University. In a stimulating lecture from the Stanford Institute for theoretical Physics *(118)*, Susskind builds on his knowledge of black holes, and

---

[25] A singularity means a point where some property is infinite. At the centre of a black hole, according to Newtonian theory, the density is infinite (because a finite mass is compressed to a zero volume). Hence it is a singularity. Similarly, if you extrapolate the properties of the universe to the instant of the Big Bang, you will find that both density and temperature go to infinity and so that also is a singularity. As yet there is no theory of quantum gravity but it is entirely possible that singularities may be avoided with a theory of quantum gravity **(115)**.

[26] Put simply the **Holographic Principle** states that everything that occurs in a space can be explained in terms of information that's somehow stored on the surface of that space. This principle is becoming increasingly important in theoretical developments to explain the reality of our universe.

his claim that information cannot be destroyed, to propose that at the edge of the universe could be an event horizon with all of the information needed to create our universe. As in the case of a black hole the event horizon is two dimensional but it transmits our world as a three dimensional hologram. Obviously I would need to undertake further extensive study before I might feel equipped to offer an informed opinion on the credibility of Susskind's hypothesis but after an initial study I consider that it seems well worth further investigation. I report on this further in Chapter 6.

I hope that in this section I have made it clear that, to contribute effectively to the provision of a more reliable explanation of reality, science still has much to do. At present we are fortunate in having available a reassuring number of outstandingly talented scientists and I feel certain that we will eventually find answers to most of the problems I have raised here. However, we must not become overconfident and must realise that science has its limits. I look at some of the limitations of science in the next section.

4.4 Limitations of Science

Science certainly has its limitations but a study of the information available on the subject of science and religion shows that the limitations of science are often not understood or ignored. Obviously this can lead to serious errors of judgement and, here, I consider some limitations we must bear in mind when using science to help explain reality. I have limited myself to four cases that I have found to be particularly troublesome.

Science is continually changing

We must bear in mind that the science we currently accept as accurate and reliable, is open to major changes as we learn more. In the first sections of this chapter I recorded some of the amazing advances which have taken place since Lord Kelvin made his famous statement, "There is nothing new to be discovered in physics now." We continue to make new discoveries which change our knowledge of the universe. What science thinks is true now may not be seen to be true as things move on, and I am certain that the major advances in science we uncovered during the last century could not have been predicted in the 19th century. As we look to the future, advances using say, quantum theory, relativity and the holographic principle, will change things even further.

Scientists make their conclusions and form their theories on observed data which, although accurate and reliable, may not give a true picture. As an entertaining example of this, the leading theoretical physicist Professor Brian Greene, in a lecture titled *Is our Universe the Only Universe? (119)*, has pointed out that there will be a time, millions of years in the future, when all the stars have moved from our sky. All that we will be able to see is an apparently infinite blackness. For scientists living during this period their conclusions about the universe will be very different from those now advanced.

Science is not geared to answer 'why' questions

The topic of 'why' questions has been dealt with at some length in Chapter 3, Section 3, where I considered the following statement from John Lennox. "There are certain questions that science is not geared to answer, particularly 'why' questions that have to do with purpose as distinct from function." I then related this to his most effective, yet simple illustration which makes the use of a newly baked cake from Aunt Matilda. From this illustration, it follows that science can describe physical objects and laws but it cannot tell us why these objects exist and explain why they obey laws. It stands to reason, therefore, that, if I claim that the existence of a creator God explains why I am here, scientists cannot agree or disagree with me.

The Laws of science have no creative power

Many scientists claim that science can prove that there is no God. Obviously, I believe that science can provide solid and convincing support to aid belief in God, but, on its own it certainly cannot prove whether God exists or not. It is, therefore, disappointing that when talking about belief in God, so many intelligent people look first to science to provide the answers and great weight is put on the opinions of eminent scientists. Recently the public waited with bated breath to discover what Stephen Hawking thought about God. According to reports *(120)*, he has now come out and declared that there is no God.

In this published interview, he states, "Before we understand science, it is natural to believe that God created the universe. But now science offers a more convincing explanation."

However, John Lennox points out that offering people the choice between God and science is illogical and in his book *God and Stephen Hawking (121)* he explains why. He first considers a statement given by Professor Hawking in his book *The Grand Design* which deals with M-theory (described earlier in Section 4.3). The statement is as follows:-

"M-theory predicts that a great many universes were created out of nothing. Their creation does not require the intervention of some supernatural being or God. Rather these multiple universes arise naturally from physical law."

John Lennox then goes on to comment:-

"A supernatural being or god is an agent who does something. In the case of the God of the Bible, he is a personal agent. Dismissing such an agent, Hawking ascribes creative power to physical law; but physical law is not an agent. Hawking is making a classic category mistake by confusing two entirely different kinds of entity: physical law and personal agency."

Physical laws do not have creative power and

without the provision of this power through an agency the universe could not exist.

Science does not make moral judgments

I believe that it is worthwhile reiterating here that while scientific knowledge can inform our opinions and decisions it is individual people who must ultimately make moral judgements. Science can help to describe the world but it cannot make any judgements about right, wrong, good or bad. Making correct judgements on what is right and wrong will be vitally important in many of the advances being made in medicine and bioscience.

Similarly, with aesthetic judgements while we can use it to analyse brilliant symphonies and interpret artistic paintings science cannot help us judge beauty or ugliness. This is left to the individual's own aesthetic taste.

Science does not tell us how to use knowledge

Finally, science does not tell us how to use scientific knowledge. For most important scientific advances you can imagine both positive and negative ways that knowledge could be used but it is up to us to decide how to use that knowledge.

4.5 Discussion

In the opening sections of this chapter I reported on the tremendous advances provided by research in biology and the physical sciences. For biology, using the theory of evolution, we can be fairly sure of how we got here, and, through the Human Genome Project, we now have all of the genetic information needed to create a human being. I have already devoted Chapter 3 to an investigation of evolution, and I will deal with the Human Genome Project in later chapters so here I will limit my discussion to the physical sciences.

For the physical sciences we can conclude that they now provide a very different picture of reality from the one held by most scientists a century ago. The concepts of reality involving quantum mechanics and chaos theory are very different from those previously held. The "solid" and "mechanical" picture of physical reality, provided by Newtonian physics, has been replaced by the bizarre and uncertain reality of quantum physics and from research using chaos theory Davies and Gribbin *(41)* can now claim that reality of a clockwork Newtonian universe filled with "things" made from inert matter has been replaced by a universe filled with "systems". They further claim that "the old machine vocabulary of science is giving way to a language more reminiscent of biology than physics".

Rupert Sheldrake *(29)* also considers that living organisms provide better metaphors for organised systems than machines and he concludes: - "the entire universe is more like a growing, developing organism than a machine slowly running out of steam".

However, we still have much to learn. There are many pressing problems, and I mentioned a few in Section 4.3. For my study on the nature of reality, the two which give me most concern are: firstly, the lack of a single theory which unites the laws of physics on the large and small scales and secondly, the need for a truly satisfactory interpretation of quantum physics, where I believe we may have to turn to philosophy to find the solution. If these problems were solved, science would be in a much more credible position. Further study of the Holographic Principle would also prove beneficial.

In Section 4.4, I stressed that we must be aware of the limitations of science when using it to help explain reality. At present science cannot by itself explain the nature of a reality which we are finding to be increasingly complex and surprising. It is not suited to answering 'why' questions that have to do with purpose as distinct with function although it is highly effective at answering 'how' questions. Also we have seen that science is continually changing and Polkinghorne, in his erudite little book *Quantum Theory – A Very Short Introduction (40)*, puts things clearly when he says, "Realists see the role of science to be to discover what the physical world is actually like. This is a task which will never be completely fulfilled. New physical regimes (encountered at yet higher energies for example) will always be awaiting investigation, and they may well prove to have very unexpected features in their behaviour. An honest assessment of the achievement of physics can at most claim verisimilitude (an accurate account of a wide but circumscribed range of phenomena) and not absolute truth (a total account of physical reality)."

To conclude this assessment of the limits of science I can think of nothing better than to finish with a wonderful quote from the brilliant physicist Erwin Schroedinger *(121)* who was such a major force in the development of quantum physics. "I am very astonished that the scientific picture of the real world about me is very deficient. It gives us a lot of factual information, puts all our experience in a magnificently consistent order, but it is ghastly silent about all and sundry that is near to our heart that really matters to us. It cannot tell us a word about red and blue, bitter and sweet, physical pain and physical delight; it knows nothing of beautiful and ugly, good or bad, God and eternity. Science sometimes pretends to answer questions in these domains but the answers are very often so silly that we are not inclined to take them seriously."

4.6 Key Points

For the purposes of this book the main points to be noted are as follows: -

1) Science is always developing and the tremendous advances in physics and biology have taken place over the last century giving us a very different picture of physical reality from the one

held by most scientists at the beginning of the century. We should note:-

- The "solid" and "mechanical" reality which seemed to exist at the beginning of last century has been replaced by the mysterious, nebulous and uncertain reality of quantum mechanics.

- Recent research using chaos theory is tending to show that the reality of a "clockwork" Newtonian universe filled with "things" made from inert matter has been replaced by a universe filled with "systems".

- It has been claimed that living organisms provide better metaphors for these "systems" and the universe is more like a growing developing organism than a machine.

- Due to major advances in biology, evolution has revolutionised our thinking on human development and we now have the complete set of genetic information needed for the creation of a human being.

2) For science to produce a more reliable explanation of reality further clarification is needed and the following should be considered:

- The physical sciences have made tremendous advances during the past century but this should not allow us to become overconfident and be led into adopting a materialist philosophy.

REALITY, SCIENCE AND THE SUPERNATURAL

- As scientists we cannot be completely confident until a unifying "Theory of everything" has been produced.

- An acceptable explanation of quantum mechanics must be found even if we need to turn to philosophy for assistance.

- Priority should be given to undertaking further research on dark energy and further investigation of the Holographic Principle.

3) Science can be used to solve an amazing range of problems and largely due to recent spectacular advances we have become almost totally dependent on it. It can prove to be an immense power for good.

However, it has its limits and these must be realised if gross errors are to be avoided. Some of these limits are as follows: -

- Scientific laws have no creative power – a law is not an agency.

- Science is not good at answering "why?" questions that have to do with purpose as distinct from function.

- Science cannot make moral or aesthetic judgements.

- Science does not tell us how to use the knowledge it provides.

CHAPTER 5

Signposts from Science to Reality

Men became scientific because they expected law in nature and they expected law in nature because they believed in a lawgiver.

*- **C. S. Lewis** (Novelist and Oxford Scholar, 1898-1963)*

The really amazing thing is not that life on earth is balanced on a knife edge, but that the entire universe is balanced on a knife-edge ,and would be total chaos if any of the natural 'constants' were off even slightly. You see, even if you dismiss man as a chance happening, the fact remains that the universe seems unreasonably suited to the existence of life --- almost contrived---- you might say a 'put up job'.

*- **Paul Davies** (Professor of Theoretical Physics, University of Adelaide)*

It is a happy day for the world. It is humbling for me, and awe-inspiring, to realise that we have caught the first glimpse of our own instruction book, previously known only to God.

- *Francis S. Collins* *(Director of the Human Genome Project)*

The rational intelligibility of the universe points to the existence of a Mind that was responsible both for the universe and for our minds.

- *John Lennox* *(Professor of Mathematics, University on Oxford.)*

5.1 Introduction

In previous chapters I have paid particular attention to the limitations of the physical sciences when they are called on to help explain reality. However, in this chapter I contend that science can provide strong supporting evidence in the form of a number of intriguing clues which act as signposts pointing to the reality of a creator God. These clues concern the reality of our universe, our physical bodies and our minds.

First I consider the order of the universe and contend that there is a strong argument that the physical laws are the work of a creator God. Then I concentrate on what has become known as the *Goldilocks effect* where, like the porridge in the tale of "Goldilocks and the Three Bears", the universe appears to be 'just right' for life. I discuss why this should be the case.

Next I consider clues of direct concern to the composition of human beings. I deal with evidence on what Francis Collins, Director of the Human Genome project, terms "the language of God" *(7)*. He uses this term when explaining the human genome – the hereditary code of life. I argue that this code is the work of a supernatural intelligence.

Finally, I examine the significance of the intelligibility of the universe to us and contend that this intelligibility provides a clear indication of the reality of the Mind of God. In the discussion which ends the chapter I argue that the clues described in this chapter help to emphasise that belief in a supernatural creative power – a creator God – makes sense.

5.2 The Order of the Universe – God's Laws at work?

There is little doubt that the doctrine of a unique creator God who is responsible for the existence and the order of the universe has played a leading role in the development of science. John Lennox, in his book *God's Undertaker (122)* states, *"*At the heart of science lies the deep conviction that the universe is orderly – where does this conviction come from?" He then goes on to produce a quote from Melvin Calvin *(123)*, a Nobel Prize winner in Biochemistry, "As I try to discover the origin of that conviction, I seem to find it in the basic notion discovered 2,000 or 3,000 years ago, and enunciated first in the Western World by the ancient Hebrews, namely that the universe is governed by a single God ....This monotheistic view seems to be the historical foundation for modern science." From this statement we can observe that, contrary to the views of many scientists, the foundation on which science stands has a strong theistic dimension.

I used these quotations in some talks I gave recently. The talks were aimed at showing that science could help to support belief in a creator God.[27] I presented several illustrations of physical laws at work and was pleased to find that my explanation of the creation of our solar system was particularly well received by the audience.

I started this explanation by noting that following the "Big Bang" things were pretty random in our universe. Then gradually over very long periods of time, order was imposed. Immense clouds of dust and gas or nebulae, were formed. Stars and planets were created from these nebulae. It was through the establishment of order in nebulae, following various laws of physics, that our solar system was created. It was particularly satisfying to explain how one simple law, *Newton's Principle of the Conservation of Angular Momentum*, played such a vital role in

---

[27] The information given in these talks helped to form the basis of a further two series of talks I gave in support of the Scientists in Scotland (SICS) project which took place during the years 2014-2016. This project was organised by SICS at the University of St Andrews with the aim of encouraging "a deeper level of conversation about faith and science throughout the Scotland".

the proceedings[28]. After witnessing such an awesome yet elegant and ordered process I became more convinced than ever that it had come about through supernatural design.

By observing and analysing the ordered processes occurring throughout the universe I contend that even atheists would agree that it looks as though this order has been achieved by design. However, the existence of the laws by themselves has failed to convince many scientists. I decided to seek further evidence to support the case for a creator God. Through extending my search to include more recent research I discovered the evidence I was looking for. In the additional examples, which follow in the next section, the laws and constants of physical science are manipulated to create conditions which allow the possibility of life. I believe that they present particularly credible evidence for the existence of a supernatural, designing intelligence.

5.3 The Goldilocks Effect

As they gain more knowledge of the universe scientists are becoming aware of numerous surprising facts concerning its uniqueness. Recently,

---

[28] To obtain a clear picture of how our solar system was formed I would recommend an excellent video, involving the well-known scientist, Brian Cox *(125)*. It provides an impressive illustration of the effectiveness of the principle of the conservation of momentum during the process of the formation of our solar system.

a number of researchers have commented that, since the laws and constants of nature are so 'finely tuned'[29] and so many "coincidences" have occurred to permit the possibility of life, the universe must have come into existence through intentional planning and intelligence. Like Goldilocks' porridge the universe appears to be "just right" for life. I now deal with three examples which, for me, point strongly to the work of a designing intelligence.

The value of entropy at the start of the universe

For one of the most amazing example of 'fine tuning' I have come across[30] we need to look at the findings of the distinguished mathematician Sir Roger Penrose *(128)* who has pointed out that the

---

[29] Over the years I have been involved in measurements in engineering research test rigs and I have come to appreciate the benefits of the fine tuning of the test rig instrumentation. Hitting the "sweet spot" through the combined operation of man and machine always gives real satisfaction particularly when it is appreciated that such fine control could not have been achieved without the guiding input of the test rig operator. Without the input of an intelligent external influence the process would not achieve its aim, so the claims that some supernatural intelligence might be involved in 'fine tuning' immediately attracted my interest.

[30] In order to explain my first example of "fine tuning" it is important that I start by defining *entropy* and then provide, for the general reader, a description of what is meant, in mathematics and physics, by a *phase-space*. Entropy can be understood as a measure of disorder and "The phase space is a multidimensional space in which each axis corresponds to one of the coordinates required to specify the state of a physical system, all of the coordinates being thus represented so that a point in space corresponds to a state of the system." *(127)*

universe must have started in a state of entropy low enough to have usable energy. He states:- "Try to imagine the phase space... of the *entire* universe. Each point in this phase space represents a different possible way that the universe might have started off. We are to picture the Creator armed with a 'pin' – which is to be placed at some point in the phase space.... Each different positioning of the pin provides a different universe. Now the accuracy that is needed for the creator's aim depends on the entropy of the universe that is thereby created. It would be relatively 'easy' to produce a high entropy universe, since there would be a large volume of phase space available for the pin to hit. But in order to start off the universe in a state of low entropy – so that there will indeed be a second law of thermodynamics[31] – the Creator must aim must aim for a much tinier volume of the phase space. How tiny would this region be, in order that a universe closely resembling the one in which we actually live would be the result?" Calculations lead to the amazing conclusion that the Creator's aim must be accurate to 1 part in ten to the power 10^{123}, that is a million billion billion billion billion billion billion billion billion billion billion billion billion billion zeros!

Penrose notes: - "Even if we were to write a zero on each separate proton and on each separate

[31] Put simply the **Second Law of Thermodynamics** states that: "Once a thermodynamic process is started you cannot return to the same energy state because there is always an increase in disorder; entropy always increases." This law is of the utmost importance in science. It cannot be violated.

neutron in the entire universe – and we could throw in all the other particles as well for good measure – we should fall far short of writing down the figure needed."

Then, commenting on what I have referred to earlier as the laws used to help govern the universe, he continues, "The precision needed to set the universe on its course is to be in no way inferior to all that extraordinary precision that we have become accustomed to in the superb dynamical equations (Newton's, Maxwell's, Einstein's) which govern the behaviour of things from moment to moment." *(124)*

As a result of these findings Penrose has gone on to conclude *(134)*, "I think I would say that the universe has a purpose. It is not there somehow by chance..." I was surprised to note that, despite making this statement on the purpose of the universe, Professor Penrose is recorded as having atheist views. He did, however complete the above statement as follows:-

"....some people, I think, take the view that the universe is just there and it runs along – it's a bit like it just sort of computes, and we somehow by accident find ourselves in this thing. But I don't think that's a very fruitful or helpful way of looking at the universe, I think that there is something much deeper about it."

I certainly have sympathy with Penrose's unease but I do not agree that the adoption of an atheistic viewpoint will help to provide a satisfactory answer. I cannot agree that, particularly once you have asserted that the universe has purpose, you can completely dismiss the existence of a creator God without providing clear evidence for this dismissal. I will expand further on my reasons for this disagreement in Chapter 6. I do contend, however, that the universe looks as though it has been designed and the example provided by Professor Penrose provides an excellent clue which points to the existence of a designer.

Production of carbon in the stars

Some 300,000 years after the *Big Bang,* hydrogen and helium atoms, formed as a result of the creation process, began to clump together into nebulae. For the next 300 million years these clouds grew to immense proportions, attracting more atoms and becoming increasingly dense and hot. Eventually the clouds became so dense and hot that they exploded in huge nuclear reactions. The hydrogen atoms then began to fuse together, creating balls of fire. The first stars were born.

All of the elements here on Earth were created in the stars. At the start of a star's life its hydrogen atoms fuse to form helium atoms. As the star grows older and the hydrogen supply is depleted the helium atoms fuse to form carbon. The carbon atoms then fuse to form oxygen. This manufacturing process continues with the elements getting progressively heavier and heavier until the star dies. Elements as heavy as iron can be manufactured using this process. Heavier elements are created in the explosions of much more massive stars, *supernovae*.

When Sir Fred Hoyle was carrying out research on how carbon came into existence in the "blast furnaces" of the stars *(124)*, his calculations showed that it is extremely difficult to explain how the stars generated enough carbon to sustain life on Earth. Hoyle found that there were numerous "fortunate" one-time occurrences which seemed to indicate purposeful "adjustments" in the natural laws in order to produce the necessary carbon. He is quoted as saying: - "A common sense interpretation of the facts suggests that a superintendent has monkeyed with the physics, as well as chemistry and biology, and that there are no blind forces worth speaking about in nature. I do not believe that any physicist who examined the evidence could fail to draw the inference that the laws of nuclear physics have been deliberately designed with regard to the consequences they produce within the stars."

Here, again, it certainly looks as though some "designer" has been at work. Indeed, commenting on the results of his research Hoyle asked, "Surely simple common sense tells us that an external intelligence is at work here?"

Dark energy and dark matter

In the year 2006, Paul Davies, one of today's most gifted and acclaimed science writers, published another of his intriguing books entitled, *The Goldilocks Enigma (126).* One of the chapters, "A Universe fit for Life", is particularly relevant to the subject matter being considered here and deals with several examples of "fine tuning". I found the section on dark energy particularly illuminating. In Chapter 4, Subsection 4.3.4 I have already mentioned dark matter and dark energy.

Dark matter plays an essential role in the shaping of the universe. It provides most of the gravitational pull needed to grow galaxies. Left to itself normal matter would prove too feeble for this to happen and life would be impossible. However, it was dark energy which provided me with perhaps the best and most important example of "fine tuning" I have come across. Dark energy is responsible for a mysterious antigravity force and is responsible for the expansion of the universe.

The future of the universe depends on it. Until fairly recently most physicists and cosmologists believed that some physical mechanism was responsible for the cancellation of the value of dark energy. However, much to their surprise, astronomers have now discovered that the influence of dark energy is not zero. Even more surprising is they have also discovered that the value of the dark energy mass density measured by astronomers is some 120 powers of ten less than the 'natural' value obtained using quantum theory. Physicists do not know why this is the case but point out that if it were 119 rather than 120 powers of ten less, the consequences would be lethal since a factor of 10 would be enough to exclude life. As Davies points out *(133)*, "A factor of 10 may seem like a large margin but one power of ten on a scale of 120 is a pretty close call."

Davies goes on to comment:-

"Logically, it is possible that the laws of physics conspire to create an almost but not quite perfect cancellation. But then it would be an extraordinary coincidence that *that* level of cancellation – 119 powers of ten, after all – just happened by chance to be what is needed to bring about a universe fit for life. How much chance can we buy on scientific explanation? One measure of what is involved can be given in terms of coin flipping: odds of 10 to the power 120 to one is like getting heads no fewer than *400 times in a row.* If the existence of life in the universe is completely independent of the big fix mechanism – if it's just a coincidence – then those are the odds against us being here. That level of flukiness seems too much to swallow."

I share the frustration shown in these words of Paul Davies.

Nevertheless, he must gain some comfort from the views of the atheist theoretical physicist and Nobel Laureate, Steven Weinberg *(130)*. Weinberg does not believe in "fine tuning" and he is responsible for such statements as, "The more the universe seems comprehensible, the more it seems pointless" *(131)*. Yet even he is surprised by dark energy and has commented that, "There is one constant that seems to be fine-tuned ---- and that is dark energy*" (132)*.

Chance or Design?

There is little doubt that on Earth here in our

universe we live in conditions that are "just right" for life. Did this Goldilocks Effect come about by chance or design? I believe that design is the answer and in this chapter I have presented three examples and I contend that each example clearly signposts a "designer". For me, the enormous odds against "chance" make the choice of "design" a safe bet. Also, unlike Steven Weinberg whom I quoted earlier, I have discovered that there are numerous other examples of this "fine tuning", such as those involving the four forces of nature, dealt with in physics – gravitational, electromagnetic, strong nuclear force and weak nuclear force. Changing the strength of any one of them, even by a small amount, could render our universe sterile. An article by the well-respected physicist Gerald Schroeder, *The Fine Tuning of the Universe (124)*, provides some useful source material. Dr Schroeder contends that most scientists believe that the universe is extremely finely tuned and even many of those who do not believe the *Anthropic Principle*[32] still conclude that the universe is "too contrived" to be a chance event. Nobel Prize winning physicist Arno Penzias *(135)* sums things up as follows: -

"Astronomy leads us to a unique event, a universe which was created out of nothing, one with very delicate balance needed to provide exactly the conditions required to permit life, and one which has an underlying (one might say 'supernatural') plan."

Before ending this section we should be aware

---

[32] The **Anthropic Principle** states that, "the observable universe has a structure which permits the existence of observers."

that, in order to explain how humans arrived on the scene, many scientists contend that a solution involving the 'multiverse' provides the answer. They argue that if there is an infinite number of universes then there must at least be one suited for human life. While I accept that the 'multiverse' explanation might provide an answer I tend to agree with Sir John Polkinghorne who suggests: -

"A possible explanation of equal intellectual respectability – and to my mind greater economy and elegance – would be that this one world is the way it is, because it is the creation of the will of a Creator who proposes that it should be so."*(136)*

It should also be stressed that the 'multiverse' explanation does not exclude the possibility of design. There is no reason why a 'multiverse' solution cannot be considered to be the work of a 'designer'. I believe that most scientists have a preference for the 'multiverse' solution because it appears to involve only science and we scientists are, naturally, much more comfortable in our own specialist area. Yet, in considering these options we should take careful note that we are leaving science behind and entering the realm of philosophy. The supernatural intelligence option, obviously, involves philosophy and I consider that the second option, the 'multiverse', also deals with metaphysics. For support for my argument on the multiverse I again turn to quantum theorist Sir John Polkinghorne *(136)* who states: "Let us recognise these speculations for what they are. They are not physics but in the strictest sense, metaphysics. There is no purely scientific reason to believe in an ensemble of universes. By construction these other worlds are unknowable to us."

I find that scientists, in general, have increasing confidence in their own subject areas and little respect for philosophy. In previous chapters I have commented on Stephen Hawking's view *(35)*, "... philosophy is dead. Philosophy has not kept up with developments in science". While in an article published in *The Guardian* newspaper in September 2012 *(137)*, the philosopher Julian Baggini worries that scientists are becoming increasingly determined to stamp their mark on other disciplines. In a discussion with Lawrence Krauss, Baggini confronts the theoretical physicist with, what he terms 'mission creep',[33] among his fellow scientists. As part of his concluding statement in what proves to be an enlightening discussion he says, "... philosophy needs to accept it may one day be made redundant. But science also has to accept there may be limits to its reach." He continues by saying that he is sceptical that human behaviour can ever be explained by physics and biology alone. Then he goes on to display the more tolerant and open-minded attitude that I find is prevalent in philosophers involved in the 'Science v Philosophy' argument, and astutely concludes to Professor Krauss, "I am happy for physicists to have a go.

---

[33] *Mission Creep,* can be defined as "the fact of doing a much larger job for a longer time than was originally expected, especially in a military operation."*(138)* Mission creep is often considered unwise due to the dangerous path of each success leading to more ambitious attempts, only stopping when a final, often catastrophic failure occurs.

But, until they succeed, I think that they should refrain from making any claims that the only real questions are scientific questions and the rest is noise. If that were true, wouldn't this conversation just be noise too?"

For questions involving science and religion I tend to give particular weight to the views of researchers who are experts in both fields. John Polkinghorne is one of the few scientists involved in the science/religion discussion who is exceptionally well informed. He is a leading theoretical physicist and a Fellow of the Royal Society. He is also a theologian and Anglican priest and, as you will have noted from my earlier comments in this chapter, I much appreciate his clear, informed and concise style of writing. In Sir John's assessment of the multiverse, as well as suggesting "how" the universe was created, he is also providing an answer to "why" the universe was created. While knowing "how" is of major importance, the question I believe we need to ask is, "Why was the universe created?" Here, since this question refers to purpose as distinct from function again science by itself has no convincing answer. The multiverse explanation does not answer the question "why?" and I find it very difficult to believe that our wonderful and finely tuned universe exists by some accident. However, the suggestion that a creator God took the trouble to undertake all of fine tuning we have considered here is, I find, rather convincing.

5.4 The Human Genome Project

The Language of God

In Chapter 4 Subsection 4.2.1 I mentioned the immense importance of discovery of DNA and described the Human Genome Project (HGP) which emerged as a result of this discovery. This project revealed an amazing code carrying with it all of the instructions for building a human being. On the day when this successful completion of the HGP was announced to the world Dr Francis Collins, the Director of the HGP, commented, "It's a happy day for the world. It is humbling for me, and awe inspiring, to realise that we have caught the first glimpse of our own instruction book known only to God."*(139)*

Many scientists, however, do not agree that the human genome is the creation of a "higher intelligence". Richard Dawkins for instance considers that, "Biology is the study of complicated things that have the appearance of having been designed with a purpose" and dismisses this comment by Francis Collins, claiming random mutations in the DNA which makes up the genome, have caused its development. These mutations certainly do appear to be random and Professor Dawkins can put forward a passionate and informed case to support his assertion that Darwinian natural selection works without design. His efforts have had a major influence in convincing an overwhelming majority of scientists in the validity of an evolution process which needs no input from God.

On the other hand Dr Collins has firm support for his belief from the Director of the Discovery Institute's Centre for Science and Culture in Seattle USA, Dr Stephen Meyer, who states:-

"When we find information in the cell (the language of DNA), this is not something that Darwinian evolution… can explain. But we do have an explanation that is known to produce information and that explanation is intelligence: conscious activity." Strongly supporting this assertion of a higher intelligence, John Lennox also states, "The moment we see text with meaning – and it's a code remember – we infer upwards to intelligence instantly." *(140)*

Let me again remind you of some facts about this amazing code.

a) The human genome consists of all of the DNA of our species – the hereditary *Code of Life.*

b) It is written in a strange cryptographic four letter code.

c) A live reading of the code at a rate of one letter per second would take 31 years even if reading day and night.

d) Printing these letters out in regular font size on normal bond paper and binding them together would result in a tower the height of the Washington Monument (555ft).

I just could not believe that this "awe inspiring" instruction book was simply the result of chance. In addition the convincing arguments in the literature, particularly in books from Francis Collins, *The Language of God (7)* and John Lennox *God's Undertaker (5)* struck a chord with me, although I knew I was going against the opinion of most scientists. *Would God devise a method which relied on chance?*

While I had made this decision, the fact that random mutations in the DNA appeared to be responsible for the development of the human genome still troubled me. Why would God choose a method which relied on chance? I searched the internet for relevant literature and came across an erudite little paper entitled, *Chance from a Theistic Perspective (141)* by Professor Loren Haarsma, from the Faculty of Physics and Astronomy at Calvin College in Michigan in the USA. To reach his conclusions Haarsma utilised information from a book by John Polkinghorne titled *Science and Providence. God's interaction with the world (142)*. I wasted no time in purchasing a copy of Sir John's book. I refer to this later.

In his paper *(141)* Professor Haarsma points out that, "The use of the term "chance" in any scientific theory is not strictly a statement about *causation* (or lack of causation); rather, it is a statement about lack of *knowledge* about causation." He goes on to note that "In evolutionary biology, a "chance" event is simply an event which is not caused by the organism itself, and which we could not have predicted given our limited knowledge of the initial conditions which affects the organism's survival (e.g. a natural disaster) or its genetic information (e.g. a mutation). "Chance" in evolution, or any other scientific theory, is a semi-quantitative statement about our ignorance - our lack of precise knowledge of the initial conditions, or our lack of understanding how a particular final state is selected."

"Random" events such as genetic mutations take place within systems of natural laws which both constrain choice and respond to the choice made and according to Haarsma, "genetic mutations can be thought of as small-step explorations of large-dimensional "genomic phase space" which was also designed by the Creator. We should also note that to "tame" chance means to break down the very improbable into less improbable small components arranged in series.

In his book, *One World: The interaction of Science and Theology* *(142)*, in a chapter entitled, *Providence*, John Polkinghorne states:-

"The world's freedom to become, and God's and our freedom to act within its unfolding process, derive from the flexibility resulting from the unpredictable sensitivity of response enjoyed by complex dynamical systems. A crude shorthand for the scientific account of one aspect of these matters is to refer to the interplay of chance and necessity. Necessity is the regular ground of possibility, expressed in scientific law. Chance in this context, is the means for the exploration and realisation of inherent possibility, through continually changing (and therefore at any time contingent) individual circumstances. It is important to realise that chance is being used in this 'tame' sense, meaning the shuffling operations by which what is potential is made actual. It is not a synonym for chaotic randomness, nor does it signify it as a lucky fluke."

Here Polkinghorne has described how modern understandings of "chaos" (see Chapter 2 Section 2.4), "allow the possibility for God to affect the outcomes of stochastic[34] processes without contravening the ordinary laws of nature and without necessarily "pulling invisible deterministic strings" during every chance event." Finally he states, "It is from this inter-relationship (between chance and certainty) that order rises out of chaos, as we see exemplified in the behaviour of dissipative systems which converge on to predictable limit cycles, approached along contingent paths… To acknowledge a role for tame chance is not in the least to deny the possibility that there is a divinely ordained general direction in which the process of the world is moving however contingent detailed aspects of that progression (such as the number of human toes) might be."

So I had received the answer to my question, raised at the start of this section, "would God choose a method which relied on chance?"

I also uncovered further backing for my contention that science can help to support belief in God when on p47 of *Science and Providence* I read Polkinghorne's following statement, "… I am still deeply impressed by the anthropic potentiality of the laws of nature which enable the small-step explorations of tamed chance to result in systems of

---

[34] A stochastic process has a random probability distribution that may be analysed statistically but may not be predicted precisely.

such wonderful complexity as ourselves. It would not happen in "any old world." That the universe is capable of such fruitfulness speaks to me of divine purpose expressed in the given structure of the world."

He also emphasises that the universe is not a universe of clockwork determinism. This fact again proves of substantial support to the arguments I have already put forward in Chapter 2 of this book.

5.5 The Intelligibility of the Universe

Most scientists would agree that the outstanding scientific genius of the 20th century was Dr Albert Einstein. In the early half of the century this truly exceptional man used his towering intellect and quite astounding insight to revolutionise scientific thought and the way we comprehend the universe. To assist me with this study I felt that it would be particularly helpful to be aware of his views on the intelligibility of the universe, but first I wanted to be clear about his opinions on religion.

Over the years conflicting reports have appeared concerning Dr Einstein's religious views. So I decided to seek clarification and my initial search centred on a set of papers on religion and science that appeared in the *New York Times Magazine* and various other publications around the 1930s to the 1950s. Copies of some the papers may be found in *Religion and Science (143)* and listed as *Religion and Science, Science and Religion 1 and 2 and*

Religion and Science: Irreconcilable? They showed that, while Einstein strongly rejected belief in a personal God and the established religions, he did admit to having "cosmic religious feeling" *(144)* and in a communication of 1929 *(145)* he stated: - "I believe in Spinoza's God, Who reveals Himself in the lawful harmony of the world not in a God Who concerns Himself with the fate and the doings of mankind." The famous 17th-century Dutch Philosopher Baruch Spinoza *(146)* believed that God exists but is abstract and impersonal. He considered that all Nature was one Reality and that God and Nature are two names for the same reality. I was impressed by the sincerity of Albert Einstein's writing on science and religion. He came over to me as a man with strong morals as well as an intense love of science.

The fact that the universe was intelligible to humans astounded Albert Einstein and he is quoted as saying: - "The eternal mystery of the world is its comprehensibility ...The fact that it is comprehensible is a miracle." He also states, "...*a priori* one should expect a chaotic world, which cannot be grasped by the mind in any way..." *(146)*.

For me the rational intelligibility of the universe provides the foundation on which the scientific method rests. I believe that the order that we can perceive in the universe is achieved through natural laws and, if these laws were not both rational and intelligible, scientific knowledge would be impossible. I also believe that the rational intelligibility of these laws supplies a vital clue to the reality of God. To illustrate my argument, in my own area of science I find the Thermodynamic Laws particularly helpful.

There are three Laws of Thermodynamics. They form the bedrock of the subject and they apply in all thermal processes. All of the complex computations which can be involved in these processes are reliant on them and these computations lead to a satisfying and often exciting appreciation of this branch of science. Of the three laws the *Second Law* is the most interesting and a corollary of this law leads us to the *thermodynamic temperature scale* and also brings us to an understanding of the property *entropy* which, as we have seen earlier in this chapter, has major significance when considering the order of the universe.

In my own research I relied on the fundamental thermodynamic laws when measuring the *transport properties* of gases as well as later projects on the measurement of sonic and supersonic gas flows, heat transfer and fluid flow and biotechnology. Over the years I have gained a rewarding insight into their significance and as my level of their comprehensibility has deepened I have become increasingly convinced that the amazing subtlety and reliability of these thermodynamic laws did not come about by accident. I believe that their rational intelligibility points to a supernatural intelligence and I believe I have gained an insight into an intelligence I can identify with. I firmly believe, as stated by John Polkingthorne *(147)*:- "It is *intelligibility* (rather than objectivity) that is the clue to reality – a conviction, incidentally, that is consonant with a metaphysical tradition stemming from the thought of Thomas Aquinas."

I have found strong support for my views on intelligibility in the book *God's Undertaker; Has Science Buried God.* In chapter 3, I find that I fully agree with the author, John Lennox *(148)*, when he contends, "The very concept of the intelligibility of the universe presupposes the existence of a rationality capable of recognizing that intelligibility."

5.6 Discussion

In this chapter my main purpose was to present a selection of clues, provided by scientific disciplines, that signpost the existence of a supernatural power and intelligence. These clues were selected since they dealt separately with aspects of realities of key interest to my project – the universe, the physical human body and the human mind. The first clue deals with the order of the universe and the "Goldilocks Enigma", the second clue examines the "Language of God", and the final clue deals with the intelligibility of the universe. I now present my main conclusions for each of them separately beginning with the order of the universe and the Goldilocks effect.

At the start of this chapter I considered the significance of the order of our universe and the natural laws which govern its operation. I claimed that at the heart of science lies the deep conviction that the universe is orderly. I felt sure that most scientists would agree with this. I also pointed out that the foundation on which science stands has a strong theistic foundation but while I felt sure that most observers would be happy to concede that it looks as though order has been achieved by design, they would not be convinced that the universe is the result of input from a creator God. Further evidence was needed. I believe that through the discovery of the Goldilocks effect we have found the extra evidence to tilt the balance in favour of a creator God.

Three examples of phenomena which contribute to the Goldilocks effect were considered under the headings, *Dark energy*, *Entropy at the start of the universe,* and *Production of carbon in the stars.* For all three I asked the question, why do they exist? Is there a reason for their existence or did they come about simply by chance? I contend that "chance" is not a likely option and the evidence for design easily wins the day. I find that the evidence emerging from the effects of recently discovered dark energy is particularly supportive of my case. Further, taking into account the fact that there are several other examples of "design" *(124)*, I conclude that there is strong evidence that our world and our universe have been designed. I close my conclusions on this first clue by reminding you of the following quote from Paul Davies *(148)*:-

"The really amazing thing is not that life on earth is balanced on a knife edge, but that the entire universe is balanced on a knife-edge, and would be total chaos if any of the natural 'constants' were off even slightly. You see, even if you dismiss man as a chance happening, the fact remains that the universe

seems unreasonably suited to the existence of life --
- almost contrived---- you might say a 'put up job'."

My second clue, presented under the heading,
The Human Genome Project, shows clearly that you
cannot convincingly "dismiss man as a chance
happening". For this clue I reported on the amazing
success the Human Genome Project (HGP) which
emerged as a result of the discovery of DNA. The
HGP revealed an amazing code carrying with it all
of the instructions for building a human being. On
the day when this successful completion of the HGP
was announced to the world Dr Francis Collins, the
Director of the HGP, remarked that "we have
caught the first glimpse of our own instruction book
known only to God". While a number of scientists
do not agree with Dr Collins I am increasingly
convinced that the awesome text of the human
genome did come about by chance and contend that
when we see this text with meaning we infer to a
supernatural intelligence.

These two clues show strong evidence for the
work of a higher intelligence in designing human
beings and then providing a world and a universe
that is particularly suited to our existence. However,
the clincher for my belief in a supernatural power
and intelligence proved to be, the intelligibility of
the universe. I found great support for my belief in
the following quote from John Lennox *(148)*:-

"The very concept of the intelligibility of the
universe presupposes the existence of a rationality
capable of recognizing that intelligibility. Indeed
confidence that our human mental processes possess

some degree of reliability and are capable of giving us some information about the world is fundamental to any kind of study, not only science. This conviction is so central to all thinking that we cannot even question its validity without assuming it in the first place, since we have to rely on our minds in order to do the questioning. It is the bedrock belief upon which all intellectual inquiry is built."

In this chapter I consider that I have shown that science can provide strong supporting evidence in the form of a number of factual clues involving evidence emerging from *God's Laws, The Goldilocks Enigma (particularly the effects of dark energy), The Language of God and The Intelligibility of the Universe*. These clues act as signposts pointing to the reality of a creator God and they have proved to be immensely supportive to my belief in God.

5.7 Key Points

Supporting evidence, in the form of clues which act as signposts pointing to the reality of a creator God, has been presented in this chapter.

Signposts pointing to the work of an intelligent designer have been obtained from:-

- *The order of the universe* – The natural laws and order of the universe make it look as though it has been formed through the efforts of a designer.

- *The Goldilocks Enigma* – Through the Goldilocks effect the universe seems particularly well suited to the existence of life. This points to the input of a designer.

- *The Human Genome* – The hereditary *code of life*, the human genome, is written in a cryptographic four letter code. The plain fact that this has meaning points to the intelligent design of the author.

- *The intelligibility of the universe* – The reasoned intelligibility of the universe points to the existence of a Mind that is responsible for the universe and for human minds.

CHAPTER 6

Adequate Evidence

We are such things as dreams are made on.

*- **Prospero** (from William Shakespeare's play, The Tempest)*

If experience of science teaches anything, it's that the world is very strange and surprising. The many revolutions in science have certainly shown that.

*- **John Polkinghorne** (Fellow of Queens College, Cambridge University)*

"As a man who has devoted his whole life to the most clear headed science, the study of matter, I can tell you as a result of my research about atoms this much: There is no matter as such. All matter originates and exists by virtue of a force which brings the particle of an atom to vibration and holds this most minute solar system of the atom together. We must assume behind this force the existence of a conscious and intelligent mind. This

mind is the matrix of all matter."

*- **Max Planck** (Founder of quantum physics)*

*Scientists are slowly waking up to an inconvenient truth – the universe looks suspiciously like **a fix**. The issue concerns the very laws of nature themselves. For 40 years, physicists and cosmologists have been quietly collecting examples of all too convenient "coincidences" and special features in the underlying laws of the universe that seem necessary in order for life, and hence conscious beings to exist. Change any one of them and the consequences would be lethal.*

*- **Paul Davies** (physicist and writer, Professor at Arizona State University)*

6.1 Introduction

On the completion of my investigation, on "signposts" to reality, I decided that I had essentially finished my initial project. Although I planned to continue with my research, I thought that I should stop for a while to review what I had learned and reflect on the evidence I had accumulated. While I had hardly "scratched the surface" of the problems to be considered, I had learned much and, in the process, become further convinced of the reality of a supernatural power. When I started my project I stated that the central aim was to strengthen my faith in a creator God by employing science to help provide answers to the following questions: -

- How credible is the philosophy of materialism?
- Does acceptance of the theory of evolution negate belief in God?
- Is science fully equipped to answer "The God Question"?[35]
- Can science produce valid evidence to support belief in the supernatural?

I needed answers to these questions in order to settle serious doubts I was having on the strength of my faith. When I started writing this book my understanding of the physical sciences showed me that they could be used to describe a physical reality which could readily, and convincingly, be explained by materialists who saw no reason for the existence of a supernatural designer. The arguments for the validity of the theory of evolution seemed to be coming increasingly solid as new fossil records were uncovered and atheist scientists such as Richard Dawkins, were advancing science-based reasons for atheism with eloquence and self-assurance. Science was continuing to make great advances and assuming greater importance in people's lives. Many scientists claimed that science on its own could explain reality and there was no need for God. Finally, I wondered if science could

---

[35] Here "The God Question" is simply taken to mean, "Does God exist?"

produce strong supportive evidence for a supernatural "designer" of the universe. In this book I have allocated four chapters to describe, separately, how I dealt with each of these four questions. Here I summarise the progress reported in each chapter and comment on how, using the evidence uncovered in my study, I have managed to answer these questions.

In the following four sections I review the explanations I have uncovered when dealing with each of the questions listed above. Having dealt with my doubts I then go on to present evidence I have uncovered which supports belief in a creator God. I then take a glimpse at the future and deal with a revolutionary new theory which is currently being finalised. If verified, this theory will again cause a major change in the way we view everyday reality. I end the chapter with a brief note on "Science and Philosophy". I follow this with my conclusions where I summarise how I have met my project objectives and contend that the results from this brief study show that science can provide sound assistance to belief in a creator God.

6.2 Materialism – a flawed and outdated philosophy

In my experience, argument formed from scientific evidence is seldom used as the main support for believing in the existence of God. Particularly when dealing with the Christian God, it is usually the atheist, often a materialist, who calls on science to support his case. However, on considering the evidence I have presented in this book, I believe I have shown that effective use of science can successfully counteract the arguments of atheists, particularly materialists.

Small scale world – quantum mechanics

As reported in Chapter 2, the new vision of reality now emerging is stimulating but challenging to comprehend. The effects of quantum mechanics make it difficult to understand the state of things as they actually exist. From everyday experience it is very hard for us to realise that, in reality, we are not surrounded by solid objects of substance. The human body, for instance, is made of atoms which are mostly space and if all this space was taken away what is left of your body it would fit into a cube less than 0.002 centimetres on each side **(149)**. The "matter" in this cube would be composed of quantum particles which lack "substance" and, as already noted, can be thought of as discrete packets of energy with wave-like properties.

Subatomic particles, such as the electron, are quantum "particles" and can exist as both a wave and a particle. A video, describing the famous *Double Slit* experiment *(152)*, gives an excellent illustration of this showing that the electron exists as a waveform until it is observed. Then it collapses from a wave into a particle in a specific location in space and time, which is what we see as reality. This result is said to be achieved by "collapsing the wave function". If you are new to quantum physics I would recommend that you study this video. I am sure that you will be as amazed as I was the first time I saw it. I have used the video to good effect in a series of talks I have given on "Science and Religion". It is usually met with disbelief by my audience until they are reminded that tests on the validity of the theory of quantum mechanics make it, by far, the most tested theory in the history of science. It is also worth noting that the famous quantum physicist Richard Feynman is reputed to have said that if you really understand the Double Slit experiment you can understand all of quantum physics.

It should be stressed that, when physicists speak of the electron as a wave they are not talking about the sort of waves we see in the ocean. It is more like a wave of possible locations where the electron could end up as a particle when it is observed. It is a wave of possibilities. The physicist Nick Herbert, author of the book, *Quantum Reality (153)*, provides a helpful way of looking at things by stating:-

"…the world behind our back, when we are not looking and cannot observe, is always a radically ambiguous and ceaselessly flowing quantum soup. But whenever we turn round and try to see the soup our glance instantly freezes and it turns back into reality… We can never really know the true nature of the quantum universe because every time we try to observe it, it turns into matter."

The "solid" and "mechanical" reality which seemed to exist has now been replaced by the mysterious, nebulous and uncertain reality of quantum mechanics. The outstanding theoretical physicist, Brian Greene, presents a superb illustration of the world of the subatomic particle in a video entitled, *The Fabric of the Cosmos (150)*. I can thoroughly recommend this work, particularly for anyone not familiar with quantum physics. It shows how the laws of quantum physics seem to bend the rules of science, revealing a world where our three-dimensional reality may be an illusion. It clearly illustrates how objects seem to be able to appear in more than one place at a time and can move into and out of existence. An object can exist almost anywhere until it is observed. What an amazing new reality is starting to be revealed!

Self-organising systems – Chaos Theory[36]

Other results reported in Chapters 2 and 5, have introduced further alterations to our established ideas on reality. These results come from recent research on *self-organising systems* which has produced the unanticipated discoveries described by Paul Davies and John Gribbin in their book *The Matter Myth* **(41).** In Chapter 2 I have shown, in some detail, how Davies and Gribbin can report that by using chaos theory and nonlinear equations,

---

[36] Chaos theory – for an excellent illustration of a chaotic pendulum I recommend the video shown in reference *(154)*. It comes from the BBC four programme, "It's Only a Theory" where useful explanation is also provided.

scientists are demonstrating that matter can be formed into systems which produce elements of spontaneity. Davies claims that these self-organising material systems can show signs of consciousness and have played a key role in the development of the reality which can be observed today. In his Templeton Prize address of August 1995, *Physics and the Mind of God (155)*, Davies contends:- "All the richness and diversity of matter and energy we observe today has emerged since the beginning (the "Big Bang") in a long and complicated sequence of self-organising physical processes. The Laws of physics not only permit a universe to originate spontaneously, but they encourage it to complexify itself to the point where conscious beings emerge who can look back on the great cosmic drama and reflect on what it all means." Davies and Gribben also state that the old "machine" vocabulary of science is giving way to a language more reminiscent of biology than physics – adaptation, coherence, organisation and so on. The existence of conscious material systems is an exciting prospect and this claim is also supported by the noted microbiologist Rupert Sheldrake. In his book *The Science Delusion* **(25)** he considers that self-organising material systems can have a mental as well as a physical aspect.

Discussion

The descriptions, just given above, illustrate that the philosophy of materialism is false and outdated. The discoveries of quantum physics and self-organising systems clearly undermine materialist views and remove what I considered to be one of the big weapons in the atheists' armoury. One of the early giants of quantum physics, Werner Heisenberg, has commented *(55)*, "It is in quantum theory that the most fundamental changes with respect to the concept of reality have taken place, and in quantum theory in its final form the new ideas of atomic physics are concentrated and crystallized. Atomic science has turned science away from the materialistic trend it had during the nineteenth century."[37]

6.3 Theistic evolution provides the best available explanation

To resolve my second doubt I had to answer the question, "Can I accept the theory of evolution and believe in a creator God?" As described in Chapter 3, to help answer this question, I reviewed what I considered to be the three most widely accepted explanations of how humanity has reached the present stage in our development. These explanations concerned, *Young Earth Creationism, Intelligent Design* and *Evolution.*

Scientific analysis showed clearly that the views of

---

[37] It is worth noting that at least three of the early "giants" of quantum physics believed in God. They were Max Planck, Werner Heisenberg and Erwin Schrodinger. They have provided me with some wonderful quotes for this book.

the Young Earth Creationists (YEC) were not valid and could be readily dismissed. The scientific evidence against their case is overwhelming and I believe that by continuing with their present arguments the YEC are doing the Christian cause much harm.

It is clear from the scientific evidence, particularly the fossil record that, while it is not without its problems, the theory of evolution gives the best explanation of how we got here. I also discovered that acceptance of the theory of evolution does not prevent belief in the existence of a creator God. While evolution involves a random process, it is not chaotic randomness nor does it signify a lucky fluke and, as reported in Chapter 5, John Polkinghorne has described how modern understandings of "chaos" allow the possibility for God to affect the outcomes of stochastic processes without contravening the ordinary laws of nature.

It was also clear that science cannot tell us why we are here and hence cannot offer a full explanation of our development. Like Francis Collins, leader of the Human Genome Project, I decided that theistic evolution offered the best available explanation.

6.4 Is science fully equipped to answer "The God Question"?

Despite our confidence in the validity of the enormous advances in the physical sciences it must be stressed that, at present, science has its limits of explanation and here I make three points: -

Our scientific knowledge is continually changing

Sometimes these changes occur very quickly and as time advances we need to pay heed to this. Not so long ago scientists believed that the "machine" provided a credible metaphor for our universe which operated using the laws of Newtonian physics. It was often referred to as "the clockwork universe". That metaphor is no longer considered viable and the quantum world now presents us with a very different and much more perplexing and complicated picture of ultimate physical reality.

This new reality shows us that many scientific conclusions based on the scientific "certainty" of the material Newtonian universe are wrong. We should learn from this and since we know from experience that science is continually changing, we would do well to heed that we might be making similar mistakes again today. If, for instance, the theoretical concepts of the hologram that I introduce here in Section 6.7, are verified, then we can again anticipate problems.

Scientists have yet to provide a satisfactory explanation for the quantum mechanics

To explain the quantum world as ultimate reality scientists have yet to find an explanation that is universally acceptable. Several suggestions have been put forward. In Chapter 2, I have discussed this and presented what I consider to be the main options, *the Copenhagen solution and the multiverse solution.* For both cases I would contend that physicists have had to turn to philosophy to find an answer. Physicists have made brilliant use of the data and ideas which emerge from consideration of the quantum world and I greatly admire their efforts. They have, after all, used the approach an engineer would adopt. To solve a problem we engineers always start by asking, "What do the data say?" and then go on to make practical use of the results leaving the theoretical physicist to get on with his analysis. However, even engineers have to admit that, until an agreed explanation is found, the reliability of our understanding of quantum physics must be questioned

Science on its own cannot answer the "God Question"

I believe that it is particularly important to stress this fact. In Chapter 3, I dealt with the subject of "why" questions in some detail. I have explained that science is not effective at answering "why" questions that have to do with purpose as distinct from function. I contended that science can describe physical objects and laws but it cannot tell us why these objects exist and explain why they obey laws. It stands to reason, therefore, that, if I claim that the existence of a creator God explains why I am here, scientists cannot agree or disagree with me.

We cannot expect science, on its own, to provide an answer to the "God Question".

6.5 Evidence to support belief in a creator God

My search for scientific evidence of supernatural design proved fruitful and I managed to uncover several examples of supporting evidence for clues I could use as "signposts" which point to the reality of a designing Mind. I selected five clues. I contend that the first three give clear indications of this Mind at work. The next two deal with the reality of the universe and are also particularly helpful in pointing to the existence of a designer. The clue on dark energy could be included as part of the Goldilocks effect but I have dealt with it separately since I am particularly enthusiastic about this clue which throws light on a vital and almost incredible control system for the universe.

I noted that signposts could be obtained from evidence listed under the headings: - *The order of the Universe, The Human Genome, The Intelligibility of the Universe, Dark Energy,* and *The Goldilocks effect.* For each of these examples my conclusions were as follows: -

The order of the Universe
Throughout history men have believed that the

order of the universe was achieved through the input of a supernatural Mind. It is not hard to believe that careful planning was involved in the formation of the *Natural Laws* which have been used to design the ordered and inspiring universe we observe today. As our knowledge of science increases, and the wonders of our universe are further explained, I contend that these men, who believed in a supernatural "designer", were right. A careful study of the natural laws and the order of the universe certainly makes it look that a designer has been at work. The results of the Goldilocks effect considerably strengthen this view.

The Human Genome

This hereditary code of life is written in a cryptographic four letter code. The plain fact that this awesome code has meaning points to the intelligent design of the author and the leader of the Human Genome Programme, Francis Collins, has aptly called it "The Language of God". In Chapter 5 Section 5.3, I highlighted the human genome as an outstanding example in my list of "signposts". This example proved to be particularly effective in pointing to God as a Mind and I quoted Stephen Meyer as follows: -

"When we find information in the cell (the language of DNA), this is not something that Darwinian evolution… can explain. But we do have an explanation that is known to produce information and that explanation is intelligence: conscious activity."

Strongly supporting this assertion of a higher intelligence, John Lennox also states, "The moment we see text with meaning – and it's a code remember – we infer upwards to intelligence instantly" *(140)*.

The Intelligibility of the Universe

Albert Einstein thought that the comprehensibility of the universe to us was "a miracle" and, in Chapter 5, I wrote that for me the rational intelligibility of the universe provides the foundation on which the scientific method rests. I believe that the order that we can perceive in the universe is achieved through natural laws and, if these laws were not both rational and intelligible, scientific knowledge would be impossible. I also believe that the rational intelligibility of these laws supplies a vital clue to the reality of God. I have argued that the reasoned intelligibility of the universe points to the existence of a Mind that is responsible for the universe and for our minds.

Dark Energy

The controlling effect of *dark energy* provided me with the best example of fine tuning I have come across. We know virtually nothing about what scientists have theorised as dark energy but it is thought to permeate all of the universe and contribute some 70% of its mass density. Dark energy is responsible for the mysterious anti-gravity force that causes the expansion of the universe. Physicists have discovered that it is tuned to be some 120 powers of ten less than the 'natural' value and point out that if it were 119 rather than 120 powers of ten less, the consequences would be enough to exclude life. It is intriguing to note such amazing control to such great accuracy on so vital a parameter by an unknown "energy". I would contend that this is further evidence of a supernatural power at work.

The Goldilocks effect

As a result of the Goldilocks effect, described in Chapter 5, the universe seems to be particularly well suited for life, and here I have selected another two examples which indicate that a supernatural designer has been involved in making the universe so particularly well suited to us. These examples are, *the value of entropy at the start of the universe and the production of carbon in the stars.* In his book, *The Goldilocks Enigma (126)*, the theoretical physicist Paul Davies presents an erudite and engrossing study. He comments: - "Scientists are slowly waking up to an inconvenient truth – the universe looks suspiciously like a fix. The issue concerns the very laws of nature themselves. For 40 years, physicists and cosmologists have been quietly collecting examples of all too convenient "coincidences" and special features in the underlying laws of the universe that seem necessary in order for life, and hence conscious beings to exist. Change any one of them and the consequences would be lethal."

In his book *The Mind of God (155)*, Davies writes, "We (conscious beings) are truly meant to be here." However, he does not believe in a creator God and considers that conscious beings exist because, "the universe is fundamentally creative in a pervasive and continuing manner, and the laws of nature encourage matter and energy to self-[organise and self- complexify to the point that life and consciousness emerge naturally".

I fully agree with Paul Davies when he says "we are meant to be here" but I do not agree that we are here for the reasons he advances. I contend that we are not here because of the laws of nature but because of the agency behind these laws. The evidence we have just considered, particularly in the examples concerning intelligibility and the human genome, cause us to "infer upwards to intelligence, instantly." For the examples of the Goldilocks effect we should do the same.

6.6 Looking to the future – another new reality?

It has been reported that many scientists in the last century held the opinion that the metaphor of universe as a giant machine, was appropriate. When considering this metaphor, Rupert Sheldrake, in his book *The Science Delusion (29)*, advances the view that, in the light of the *Big Bang Theory*, a metaphor which describes the entire universe as a growing, developing organism is more appropriate than a machine slowly running out of steam. Also, according to Sheldrake, claims such as those given by Davies and Gribbin in their book *The Matter Myth (41)*, are changing the mindsets of many modern scientists.

As reported in Chapter 5, the mass density of the universe can be divided into 70% dark energy, 26% dark matter and only 4% of matter as we know it. So dark matter and dark energy compose 96% of the mass density of the universe. We know little about dark matter and virtually nothing about dark energy. It is theorised that dark energy permeates all of the universe. So to visualise the universe as being essentially a vast expanding "cloud" of energy does not, at first sight, seem unreasonable.

Further discussion on these ideas may be worthwhile but there is no doubt that the physical sciences, over the past century, have given us a very different picture of the universe from the one held by scientists at the beginning of last century. In this book I have discussed the advances made in quantum physics and research using chaos theory is also starting to show that the reality of a "clockwork" Newtonian universe filled with "things" made from inert matter is wrong and a universe filled with "systems" gives a better picture. These advances as well as the discoveries of dark energy and dark matter have altered my picture of physical reality. Looking to the future, in my view, there is no doubt that further major changes will take place and I close this chapter with a review of a recent theory which might well provide further major changes to our thinking.

The Holographic Universe

I have considered the *Holographic Principle* in the conclusions to Chapter 4. This principle is used in a revolutionary new theory which could assist with the explanation of reality. If verified it will cause a truly astounding change in our appreciation of what is real. Commenting on the importance of this theory, the theoretical physicist Brian Greene has stated, "If It's right, just as Newton and Einstein completely changed our picture of space, we may be on the verge of an even more dramatic revolution" *(150)*. Some years ago I heard the distinguished theoretical physicist Leonard Susskind talk about the possibility that our universe could be a hologram *(156)*, and since then, I have been intrigued by the prospect.

Most of us, at some time, have come across a hologram. It can be described as "a three-dimensional image reproduced from a (two-dimensional) pattern of interference produced by a split coherent beam of radiation (a laser)."[38] It is important to realise that a hologram is a virtual image and if you are external to it you can prove this by trying to touch it and finding there is nothing there. If, however, you are part of the hologram this will not be the case. This makes it difficult for you to realise that you are, in fact, part of a hologram. A group of scientists, among them some of the most renowned world experts in their field, now claim

---

[38] For a clear explanation of how a hologram works I recommend that you consult reference *(159)*. In recent years great improvements have been made in the production of convincingly accurate moving holograms.

that we are living in a hologram, our reality is a virtual image, an illusion that is not real.

As was the case in my introduction to quantum physics, my initial reaction to this assertion of the reality of the *Holographic Universe* was one of disbelief. However, as it is being championed by some of the finest minds currently involved in theoretical physics, I resolved to try and find out why they had come to such a surprising conclusion using mathematical models.

The development of the theory of the holographic universe came about as a result of research into *black holes*[39]. This research indicates that physical reality is two-dimensional and Brian Greene *(160)* claims that, if it is correct, "You and I and even space itself may actually be a kind of hologram. That is, everything we see and experience, everything we call our three-dimensional everyday reality, may be a projection of information stored on a thin, distant two-dimensional surface." Leonard Susskind[40] suggests that this two-dimensional surface stretches round the outer edges of our universe and acts like the surface of a black hole. From this surface our universe is projected as a sort of three-dimensional hologram.

---

[39] I have mentioned black holes in Chapter 4 Section 4.3. They are a source of great interest to today's leading scientists and the results of their research is providing major improvements in our knowledge of the universe

[40] For further information on the holographic universe I would recommend that you watch Leonard Susskind's lecture, *Leonard Susskind on the World as a Hologram (118).* This one hour lecture provides some useful insights into Susskind's arguments.

To illustrate how this claim might be possible, in his video *The Fabric of the Cosmos* *(150)*, Brian Greene uses a thought experiment to demonstrate the disconcerting properties of the surface of a black hole. In this experiment he employs his wallet as an example of a three-dimensional object. At the start of the demonstration the wallet is thrown into the entry of a black hole. As we would expect, it is immediately pulled towards the *singularity* at the heart of the black hole, never to be seen again. However, that is not the whole story. Mathematicians have discovered that the "information" needed to define the wallet and its contents can transferred to the internal surface of the black hole as it falls towards the centre. In a process which can be described mathematically, the "information" is spread on to the two-dimensional surface of the black hole and is stored there in much the same way as information is stored on a computer. So that in the end there are two copies of the wallet. A three-dimensional version which is lost in centre the black hole and a two-dimensional version on the surface. In theory the information stored on the surface can be used to recreate the wallet in three dimensions. Here it is important to note that since the space inside a black hole is no different from the space elsewhere in the universe, operation of this recreation process need not be confined to black holes. In theory the process could take place anywhere in the universe, making possible the production of a holographic universe

Much remains to be done before this theory can be progressed and verified. Many questions required to be answered, such as: -

"Where did the information stored in this two-dimensional surface come from?"

"Why is it there?"

"Is the holographic universe the result of design?"

Obviously, of particular interest to us is the answer to the question, "Can we accept that our physical bodies (and our minds?), are in reality, stored as "information" on a distant two-dimensional surface?" It will be fascinating to see how things develop. I include mention of this theory here to give an indication of the lengths to which current experts in the field will go to find answers to the perplexing problems they are facing. I also wish to illustrate how scientific views of reality are constantly changing as they seek answers to unexplained problems. Interestingly, at the start of his lecture on "the World as a Hologram" *(161)*, Susskind says that his holographic theory came about because of advice from Arthur Conan Doyle's famous detective, Sherlock Holmes, who stated, "When you have eliminated all that is impossible, whatever remains must be the truth, no matter how improbable."

6.7 Science and Philosophy

Before finally listing the conclusions of my project it is important that I comment on the benefits of forming working relationships between scientists and philosophers.

This project has proved to be extremely informative for me. I have discovered much that is new and greatly expanded my knowledge of what I can accept as real. Niels Bohr's astounding statement, "Everything we call real is made of things which cannot be regarded as real," *(59)* amazed and excited me. It introduced me to the world of quantum physics which appears to be a sort of dream world where you can forget about common sense and where objects have no substance. When we consider what is real about ourselves it seems that William Shakespeare's Hamlet got it wrong when he talked about his "solid flesh" *(157)*, but Prospero, from *The Tempest* got nearer to the truth when he stated, "We are such stuff as dreams are made on" *(158)*. Due to advances in science, the well-defined differences which appeared to exist, between the nature of the mind and the nature of the material world, no longer seem so distinct. The reality of the quantum world, the signs of consciousness in matter and the signs of an animate universe reported in earlier chapters, present increasing evidence to support my use of the metaphor of God as a Mind (see Chapter 2, Section 2.4) To fully assess the significance of this development I contend that the scientist should look to the philosopher for assistance.

However, as I have reported in Chapter 5, in my experience scientists seem reluctant to involve philosophers in any study which could involve both science and philosophy. I find this disappointing since the quantum world has yet to be fully explained and, as mentioned in Chapter 2, Chad Orzel in his book, *How to teach quantum Physics to your Dog (55)*, has stated: -"Quantum theory's effect on science goes beyond the merely practical – it forces physicists to grapple with issues of philosophy."

For me the *multiverse theory*, which is favoured by most of leading quantum physicists, is strictly metaphysics. So, like it or not, scientists have already had to turn to philosophy, and believe in the supernatural, for assistance. As a final comment it must be amusing for theists to note that, at the present time, atheist quantum physicists are being told that their observations lack common sense and, since we can never know these universes, their belief in the supernatural can never be proved!

6.8 Conclusions

My main conclusions can be listed as follows:-

1. The influences of quantum physics, and self-organising systems, undermine the arguments of the atheist materialist and show that the philosophy of materialism is seriously flawed and outdated.

2. Acceptance of the theory of evolution does not prevent belief in the existence of a creator God and I contend that theistic evolution provides the best available explanation of how humans have developed.

3. We must be careful when using science to help with the "God Question" since, at present it is

difficult to build up a reliable scientific picture of the real world. For instance:-

• *We are still waiting for a clear and undisputed explanation of quantum physics and scientists are turning to philosophy to find solutions.*

• *Some 96% of the mass density of the universe is composed of dark energy and dark matter. We know little about dark energy and virtually nothing about dark matter.*

• *It is also true that science is continually changing and if the emerging theories, involving The Holographic Universe, or Dark Matter are verified the consequences will again be game-changing.*

4. There are numerous cases where the results of science point to the existence of a creator God and particular mention has been given here to:-

• *Goldilocks Effect – particularly the fine tuning of dark energy*

• *The Human Genome*

• *The Intelligibility of the universe*

I have achieved my main aim of showing that, with care, science can be used to support belief in a creator God and strengthen my faith. I have managed to rid myself of the four main doubts which plagued me at the start of my study and I have discovered satisfying evidence which points to the existence of a creator God.

I opened the Preface to this book with a quotation from the Anglican theologian, W. H. Griffith Thomas which states:-

"(Faith) affects the whole of man's nature. It commences with the conviction of the mind based on adequate evidence; it continues in the confidence of the heart or emotions based on conviction, and it is crowned in the consent of the will, by means of which the conviction and confidence are expressed in conduct."

I believe that I have now accumulated this "adequate evidence".

APPENDIX 1

Suffering and Evolution

In my experience when non-believers are asked if they believe in a creator God they often reply with statements like, "I cannot believe in the existence of a god who allows so much cruelty to exist in the world." However, God's existence does not depend on the perceived moral aspects of His nature and, to avoid confusion, particularly when dealing with science and the God Hypothesis [41] I believe that priority should be given to answering questions on God's actuality before any reference is given to His perceived morality. Bearing this in mind, so far, I have reported on my investigation into the existence of God with hardly any reference to moral aspects of His nature. In particular, as already mentioned in Chapter 3, I have not made any attempt to answer the question, "Why does God allow so much suffering to occur in the evolution process?"

---

[41] The God hypothesis states, "a superhuman, supernatural intelligence exists who deliberately designed and created the universe and everything in it including us."

This is a question which has plagued me, and I fear many others, for some time now. However, during my project I discovered that science could come up with some helpful facts and, towards the end of my study, I came across a very rewarding paper by John Polkinghorne, entitled, *Does God Interact with his Suffering World? (98)*. The paper gives support to several of my own opinions on the reasons for suffering and deals with subjects which have been discussed in earlier chapters of this book. These subjects include: - evolution (Ch3), the unpredictable scientific world (Ch4), quantum unpredictability (Ch2), and complexity theory (Ch2). However, here I restrict my attention to evolution before adding a final section on the significance of my belief that God can be considered to be a Mind.

To clarify things on the problem of suffering in the evolution process I would contend that we need to answer two main questions as follows: -

1) Why did God select evolution for the creation process?

2) Having selected evolution why doesn't God interfere with its cruel aspects?

For question (1), Polkinghorne explains that God has to choose a path between two unacceptable extremes. On the one hand God can act as a cosmic tyrant, who causes everything to happen and on the other hand God can simply start things off and then stand back and watch. Polkinghorne is quick to stress that since our Christian God is a God of love, neither path is acceptable and Christian theology has to steer a course between these extremes. In his view it has to "speak of a God who interacts with the world but does not overrule creation". For the case of the cosmic tyrant he states, "The Christian God is a God of Love, and the God of love cannot have a creation that is simply a divine puppet theatre, of which God is the great cosmic master." He then adds, crucially in my view, that, "The gift of love always has to be the gift of some kind of appropriate freedom to the object of love; parents know that; we allow our children to grow up and be themselves, and in the same way God allows creatures to be themselves."

Polkinghorne also presents the following statement on Darwin's insights, given by the famous English clergyman and novelist, Charles Kingsley: "By bringing a world with evolution in it God has made a world in which creatures can "make themselves". God has endowed the world with very great potentiality, but the way that potentiality emerges into actuality is by the shuffling exploration of natural selection, in the course of which creatures make themselves, there by generating an astonishing three and a half billion year history, which turned the world that was already a billion years old and had bacteria in it, into a world which today contains you and me."

Polkingthorne then expresses the view that this world in which creatures make themselves is a greater world than a ready-made world would have been. He adds, "It is the gift of love that not only can creatures be themselves, in the sense that I have been trying to indicate, but they are also allowed to make themselves." In my view we have now found a reasonable explanation as to why God chose the path of evolution.

Turning now to the question, "Having selected evolution why doesn't God interfere with its cruel aspects?" Polkinghorne astutely describes the evolutionary process as "the shuffling exploration of happenstance", and points out that, "its explorations may and indeed do, bring to birth great fruitfulness, but they **inescapably** also generate ragged edges and blind alleys. **We can't have one without the other**[42]." He goes on to comment, "...the scientific insight of evolution shows us that it is not gratuitous. It is not something that if God was a bit more careful, or a bit less callous, could easily have eliminated. It is the inescapable shadow side of a world in which creatures make themselves." So here we now have another reasonable explanation of why God does not interfere with the evolution process.

Polkinghorne sums things up by concluding: -

"Quite frankly, we all tend to think that if we were in charge of creation we would have done better. We would have kept all the nice things, the sunsets the flowers and that sort of thing, and we would have got rid of all the nasty things, the disease and disaster, in the world. But the more science helps us to understand how the world actually works, the more we see that those things are intermeshed with each other. You can't tear them apart, saying "Here's the

---

[42] I have added the heavy print to emphasise the importance of the point being made.

good – keep that, there's the bad – throw that away!" It is a sort of package deal. You can't have one without the other."

Before I finish this Appendix I would like make a crucial point about human suffering. As I have already indicated I believe that God can be considered as a Mind and we form part of it. I believe that, since we are part of His being, it is an awesome act of love that, for our sake, He suffers in the evolution process with us.

APPENDIX 2

The Moral Law – The Start of My Christian Journey

In my views, DNA sequence alone, even if it is accompanied by a vast trove of data on biological function, will never explain certain special human attributes, such as the knowledge of the Moral Law and the universal search for God.

*- **Francis Collins** (Director of the Human Genome Project.)*

The ideal is there. The way to the ideal is committal to Jesus Christ; and for that committal we do not need to wait for perfect understanding; we can begin with love.

*- **William Barclay** (Professor of Divinity, Glasgow University)*

Introduction

While it was my keen interest in the interaction between science and religion that caused me to write this book, my Christian faith was not founded on a scientific law. It was founded on the *Moral Law* which came to my attention on reading the book *Mere Christianity* by C. S. Lewis *(161)*. It was acceptance of this law that led me to form my commitment to Christian love and morality as described in the *New Testament* of the *Christian Bible* and I have used Christ's teachings on morality and love, as well as St Paul's explanation of love, to form the mainstays of my faith.

On completion of my book I felt that I should add this Appendix to show what started me on my journey to Christianity. I hope that this proves of assistance to those who, like me, are constantly questioning their faith, and strengthens them in their resolve.

The Moral Law

A vital clue to both the existence and moral nature of God emerges from the *Moral Law* which is described in the ground breaking book, *The Language of God – A Scientist Presents Evidence for Belief (7)*, by Francis Collins. As reported earlier in Chapter 3, Dr Collins, one of the world's leading scientists, was the Director of the famous *Human Genome Project*. During this project he led a team of international scientists, over a period of more than ten years, to reveal the *Code of Life* which Collins *(165)* describes as: - "An amazing script carrying with it all of the instructions for building a human being." This brilliant work to map a DNA sequence, is of quite unbelievable complexity and enormous significance to human development.

I found the *Language of God* compelling reading and was very comfortable with the scientific method employed by Collins throughout the work. I was particularly intrigued when, in the first chapter, he introduced the Moral Law which states that the sense of right and wrong is an intrinsic quality of humans. Scientists, such as Francis Collins, are involved deriving and applying laws. We know our laws are true by continually verifying them against existing data. Scientific laws govern the order of our universe. My subject area, thermodynamics, is built on three laws, and while, as a scientist, I am trained to constantly question, I apply these laws with the utmost confidence. To my surprise here was Francis Collins applying the same confidence to the Moral Law.

When describing this law he states *(166)*:- "What we have here is very peculiar: the concept of right and wrong appears to be universal among all members of the human species (though its application may result in widely different outcomes). It thus seems to be a phenomenon approaching that of a law like the law of gravitation or special relativity." He then goes on to comment:- "It is the awareness of right and wrong, along with the development of language, awareness of self, and the ability to imagine the future, to which scientists generally refer when trying to enumerate the special qualities of *Homo sapiens.*"

Collins then asks:- "Is the sense of right and wrong an intrinsic quality of humans or are there other possibilities. Is it the consequence of cultural conditions or even an evolutionary by-product?" With excellent support from by C. S. Lewis in his inspirational book *Mere Christianity (161)* and some very effective argument using the selflessness of altruism, Collins, in my view, effectively argues against these possibilities, stressing that:- "Selfless altruism, presents a major challenge for the evolutionist. It cannot be accounted for by the drive of individual selfish genes to perpetuate themselves."

The Moral Law and the existence of God

When explaining the Moral Law, C. S. Lewis states, "Human beings all over the earth have the curious idea that they ought to behave in a certain way and cannot get rid of it." During the first five chapters of *Mere Christianity* he argues that the Moral Law points to the existence of God. His argument can be outlined as follows *(163)*:-

- There is a universal Moral Law.
- If there is a universal Moral Law, hence there is a Moral Law giver.
- If there is a Moral Law giver there is something beyond the universe.
- Therefore there is something beyond the universe.

Defending his argument for the existence of God against the apparent lack of material evidence for God, Lewis *(164)* responds:- "If there was a controlling power outside the universe, it could not show itself to us as one of the facts inside the universe – no more than the architect of a house could actually be a wall or staircase or fireplace in that house. The only way we could expect it to show itself would be inside ourselves as an influence or a command trying to get us to behave in a certain way. And that is just what we do find inside ourselves. Surely this ought to arouse our suspicions?"

Lewis also contends, "…You find out more about God from the Moral Law than from the universe in general since you find out more about a man by listening to his conversation than by looking at a house he has built."

The moral nature of God

When I stop and consider the Moral Law, and I realise just how often it influences and even controls my life, I become further aware of its great importance. The wonderful influences of true altruism are part of the love we are equipped to experience and contribute to the stature and dignity that makes mankind so special. I believe that the Moral Law gives us a clear insight into the nature of God who expects us to use our sense of right and wrong to select and then adhere to a moral code we believe in. Acceptance of the Moral Law proved of prime importance to me when I was forming my belief in the essence of the Christian faith.

While Lewis stresses that we should note that his initial consideration of the Moral Law is just a start and he is, "not yet within a hundred miles of the God of Christian theology" *(167)*, I became convinced by his arguments on the existence of God. I found in the Moral Law a solid foundation on which to construct my belief in the loving Christian God.

Mainstays of my faith

On this foundation I built the two mainstays of my faith. Essentially these mainstays were formed from information I obtained from the *New Testament* of the *Christian Bible* and can be grouped under the headings, "Love and Saint Paul" and "The Sermon on the Mount and the Beatitudes".

Love and St Paul

The Christian faith is built on love and this can be clearly verified by reading the *New Testament*[43] where in *Matthew ch22 v37, The Greatest Commandment* reads:-

"You must love the Lord your God with your whole heart and your whole soul and your whole mind."

Even those who do not believe in God must

---

[43] For my quotations from the Bible I have used a translation of the *New Testament* by William Barclay *(169)*.

admit that, if you replace "the Lord your God" with "Goodness", this is a request that they can wholeheartedly agree with. It is clear to me that this Commandment gives a wonderful start from which to build a moral code. I am further assured when I read the commandment given in *Mark ch12 v31*:-

"The second commandment is this; you must Love your neighbour as yourself."

Before continuing it is vital that I make clear what I mean by the word "love". In modern society the meaning has become distorted and diluted. For instance in the statements, "I love my baby," and, "I love the colour of your new car," the word "love" has quite different meanings. When a couple say, "Let's make love," in most cases they mean simply that they want to indulge in sexual intercourse. Indeed often, with modern usage it is difficult to separate "lust" from "love". What I mean by love has been clearly explained some time ago in a delightful book, *The Four Loves (169)* by C. S. Lewis. In this book the four loves are listed as, *Affection, Friendship, Eros and Charity.* Lewis shows how each of these loves can merge into one another and, more importantly from my point of view, illustrates the deceptions and distortions which can make the first three, the natural loves, dangerous without the grace of charity.

Charity can be described as, "The divine love which must be the sum and goal of all" *(169)*. It is reckoned to be the greatest of the three theological virtues. The famous and highly influential medieval theologian St Thomas Aquinas placed charity in the context of the other Christian virtues and specified its role as "the foundation or root" of them all. The word "charity" is derived from the Latin *caritas* meaning the love illustrated by the sort of "selfless altruism" attributed to the Moral Law by Francis Collins mentioned earlier in this chapter. I was also taken by the fact that Christian theologians reckon that *caritas* originates in the *will* rather than emotions. Again, according to Saint Thomas Aquinas charity is an absolute requirement for happiness. He also considered that charity has two parts: love of God and love of man which includes both love of one's neighbour and one's self. In his First Letter to The Corinthians, chapter 13, Saint Paul provides the most beautiful and moving description of Christian charity. Here I would recommend the use of the King James Version of the Bible of St Paul's description since it uses "charity" [44] in place of "love". It is important to eliminate any of the dubiety, which exists in the use of the word "love", from Saint Paul's moving description.

For a large number of Christians this chapter of Corinthians is the most wonderful chapter in the whole of the New Testament and it is sometimes referred to as *The Hymn of Love.* For a Lowland Scot like me its words are splendidly enhanced by the use of my native tongue in William Lorimer's translation, *The New Testament in Scots (170).*

---

[44] As in the more modern translations of the New Testament, when, in this book, I use the word "love" without qualification, I mean charity.

The Sermon on the Mount and the Beatitudes

It is agreed by most that the words, spoken by Jesus during the sermon, provide the essence of Christian faith and life. It is a challenging, deeply absorbing and meaningful text dealing with a range of topics which can be grouped under some twenty headings such as love for enemies, giving to the needy, prayer and fasting. The sermon is preceded by a section (*Matthew ch5 v1-11*) entitled *The Beatitudes* and this section has had particular attention from me ever since some time ago I bought a copy of *The Plain Man Looks at the Beatitudes (171)* written by Professor William Barclay in 1963. The book cost the princely sum of two shillings and six pence or half a crown.

William Barclay was born in Wick in the far north of Scotland on the 5th of December 1907. He died In Glasgow on the 24th of January 1978. He was educated at Dalziel High School in the town Motherwell, a close neighbour of the town of Hamilton, where I have spent most of my life, here in Lanarkshire. Like me, he attended Glasgow University. He became a leading figure and popular broadcaster for the Church of Scotland particularly during the 1960s and I can confirm that he was held in great esteem by his many listeners. He was an on outstanding scholar, published over 30 books and proved a very distinguished Professor of Divinity and Biblical Criticism at the University. While the two of his books which had most attraction for me were *The Mind of Saint Paul (172)* and *The Plain Man looks at the Beatitudes (171)*, his greatest success was the *Daily Study Bible (173)* which provided a set of seventeen commentaries on the New Testament. In this work William Barclay goes verse by verse, through his own translation of the Bible including every possible interpretation he knew and providing relevant background information. The study was aimed at the layman and Professor Barclay's main aim was, "Making the best possible scholarship available to the average reader."

He certainly achieved this and since we shared such a similar cultural background it is not surprising that I found that I appreciated his work and applauded his sometimes unorthodox views. I was particularly taken by his belief in *Universal Salvation* where he put forward the view, "I am a convinced Universalist. I believe that in the end all men will be gathered into the love of God." His book on the Beatitudes is a compilation of the lectures Professor Barclay gave to students at Trinity College in the University of Glasgow and with characteristic honesty he dedicates the work as follows, "To the students of Trinity College, past and present, who have already heard what is in this book."

The version of the Beatitudes given in *Matthew ch5* of the King James Bible has always held a special affection for me although there are more up-to-date translations which provide immediately clearer meaning for the reader. Working out the implications of the paradoxes presented in the Beatitudes enables the reader to discover "the technique of being a Christian" and in his book Professor Barclay provides explanations which I found extensive, enlightened and informed.

At the beginning of his book he states, "For most people the Sermon on the Mount is the essence of the Christian faith and Life; and equally for most people the Beatitudes are the essence of the Sermon on the Mount. It is therefore not too much to say that the Beatitudes are the essence of the essence of the Christian Way of life."

Following the Christian Ethic

I agree with William Barclay and I strive to live according to the teachings of the Sermon on the Mount. I find that most people I meet agree with the Christian ethic but most also consider that they can have this ethic without being bothered about the religion and thinking about Jesus and the claims he made on men. Before I started on this reassessment of my faith I had some sympathy with this point of view, but here William Barclay is emphatic in his rejection of this disregard for the importance of religion and states:-

"...the Christian ethic is only possible for the committed Christian. The proof of this statement is obvious. The world has had the Sermon on the Mount and the Christian ethic clear before it for almost two thousand years and it is no nearer to achieving it and working it out in practice. It still remains a dream and a vision."

And this dreadful decline is taking place in a society where, as I have already said, most people I meet claim that Christian morals ought to be the order of the day. Here William Barclay has some strong words:-

"There is in this world an obvious difference between *ought* and *can.* There is all the difference in the world between what man ought to do in theory and what he can do in fact. It may be perfectly correct to say to a fat, flabby, out of condition middle aged man that he *ought* to be able to run the hundred yards race in ten seconds, but the plain fact is he cannot do so. When we think of a world in terms of men and women living and acting towards each other on the basis of the Sermon on the Mount, the whole dream is complete impossibility without that committal to Jesus Christ from which the ability to live this kind of life springs. Only he who gave us his commands can enable us to obey these commands."

From Professor Barclay's words it is clear that "the Christian ethic can never be divorced from commitment to Jesus Christ" but achieving the level of commitment required here posed a real problem for me. There is much about Christianity which I cannot fully accept or understand and, in particular for me, acceptance of the validity of much of the Old Testament is, to say the least, difficult. However, here again William Barclay had some helpful advice:-

"The problems clear up as we go ahead (with our commitment to Christianity). The man who waits to understand everything will wait forever. We must begin with what we know and as we go on we will understand more and more. The ideal is there. The way to the ideal is committal to Jesus Christ; and for that committal we do not need to wait for perfect understanding; we can begin with love."

I have enthusiastically heeded William Barclay's advice.

REFERENCES

Preface

1. Richard Dawkins, *The God Delusion,* London, Transworld Publishers, 2006

2. Richard Dawkins, *The Greatest Show on Earth,* New York, Simon & Schuster, Inc. 2009

3. Alister McGrath, *Dawkins God,* John Wiley and Sons Limited, 2013

4. Alister McGrath and Joanna Collicut McGrath, *The Dawkins Delusion,* London, SPCK, 2007

5. John Lennox, *God's Undertaker. Has Science Buried God? ,* Oxford, Lion Hudson plc, 2007

6. Keith Ward, *Why There Almost Certainly is a God,* Oxford, Lion Hudson plc, 2008

7. Francis Collins, *The Language of God,* London, Simpson and Schuster UK Ltd., 2007

8. Google*, All about the Human Genome Project (HGP),* 2015

9. Google, *Seth Lloyd,* 2015

10. Neil Munro, *Para Handy,* Edinburgh, Birlinn Ltd., 1992

11. Google, *Engineering quotes at CV Engineering*, 2015

12. Google, *N W Dougherty quotes – the Quotations page 2015*

13. Google, *Quotes by Carl Mitcham | A-Z quotes,* 2015

14. *Dawkins' God, p86*

Chapter 1

15. John Polkinghorne, *Quantum Physics and Theology. An unexpected Kinship,* London, SPCK, 2007

16. *Why There Almost Certainly is a God*, p18

17. *Why There Almost Certainly is a God*, p26

18. C. S. Lewis, *Mere Christianity,* Glasgow, Collins, 1981

19. Max Planck, *The Nature of Matter,* speech at Florence, Italy (1944) (from archive zur Geschichtemder Max Planck-Gesellschaft, Abt. Va, Rep. 11 Planck, Nr. 1797)

20. Daniel Dennett, *Consciousness Explained,* London, Penguin Books, 1993

21. Christopher Hitchens, *The Portable Atheist,* Philadelphia, Da Capo Press, 2007

22. Peter Atkins, *On Being,* Oxford, Oxford University Press, 2011

23. Stephen Hawking and Leonard Mlodinow, *The Grand Design,* London, Transworld Publishers, 2010

24. Keith Ward, *The Big Questions in Science and Religion,* PA, USA, The Templeton Foundation Press, 2008

25. Google, *Rene Descartes Biography*, 2015

26. C. S. Lewis, *Miracles,* London, Collins, 2012

27. *Why There Almost Certainly is a God,* p22

28. Google, *Thomas Hobbes Biography*, 2015

29. Rupert Sheldrake, *The Science Delusion,* Croydon, Coronet, 2013

30. Richard Dawkins, *The Selfish Gene,* 2nd edition, Oxford University Press, 1989, p198

31. *The Selfish Gene,* p198

32. *The God Delusion,* p51

33. *Why There Almost Certainly is a God,* p25

34. *The God Delusion,* p135

35. *The Grand Design,* p13

36. Karen Armstrong and Richard Dawkins, *Man vs. God,* Wall Street Journal, life/essays, Europe Edition, September, 2015

37. *The Grand Design,* p227

38. Paul Davies, *The Goldilocks Enigma,* London Penguin Books, 2007

39. Walter Isaacson, *Einstein: His Life and His Universe*, New York, Simon and Schuster, 2007, p462

40. John Polkinghorne, *Quantum Theory, A very short introduction,* Oxford, Oxford University Press, 1974, p86

41. P. Davies and J. Gribbin, *The Matter Myth,* New York, Simon and Schuster Inc., 2007

42. Isaac Newton, *Principia Mathematica,* 1687

43. *The Language of God,* p199

Chapter 2

44. *The Concise Oxford Dictionary,* Oxford, Oxford University Press, 1974.

45. Google, *Dictionary of Physics – Oxford Reference*, 2015

46. Albert Einstein, *Relativity: The Special and General Theory,* New York, Bartleby.com, 2000

47. Google, *Fermilab*, 2015

48. Google, *Hadron collider*, 2015

49. *Why There Almost Certainly is a God*, p15

50. Max Planck, "On an Improvement for Wien's Equation for the Spectrum", *The Old Quantum Theory,* Pergamon Press, 1967 (translated by D Haar.)

51. Google, *Einstein photoelectric effect paper 1905*

52. Google, *Heisenberg's Uncertainty Principle – Science – How stuff works*, 2015

53. John Polkinghorne, *Quantum Theory – a very short introduction,* London Oxford University Press, 2002

54. Chad Orzel, *How to Teach Quantum Physics to Your Dog,* Oxford, Oneworld Publications, 2010

55. *How to Teach Quantum Physics to Your Dog,* p5

56. John Polkinghorne, *Reason and Reality,* London, SPCK, 1991

57. Google, *The Copenhagen Interpretation,* 2015

58. Google, *The many worlds interpretation,* 2015

59. Google, *Neils Bohr quotes, brainyquote,* 2015

60. Google, *Virtual Particle 2015*

61. Brian Greene, *The Elegant Universe,* London, Johnathan Cape, 1999

62. *Oxford Dictionary of Philosophy,* Oxford University Press, 2008

63. Google, *Andrew Zimmerman Jones - Can string theory be tested?,* 2015

64. *The Matter Myth,* pp 10-29

65. Google, *What is Chaos Theory? – Fractal Foundation,* 2015

66. *The Matter Myth,* p36

67. *The Matter Myth, pp 41-42*

68. *The Matter Myth pp 61-62*

69. Baruch Spinoza, *Ethics,* London, Penguin Classics, 2004

70. Google, *Gottfried Leibniz,* 2015

71. Keith Ward, *God and the Philosophers*, London, Fortress Press edition, 2009, pp 74-75

72. Galen Strawson, *Realistic Monism, why Physicalism Entails Parapsychism,* The Journal of Consciousness, Studies, 13, 2006

73. *The Science Delusion*, p55

74. Lee Smolin, *The Trouble with Physics,* UK, Penguin, 2006

75. *Quantum Theory – a very short introduction*, p85

76. *The Matter Myth,* p14

77. *The Science Delusion,* p128

Chapter 3

78. Google, *education Berkley edu evolution 101,* 2015

79. Charles Darwin, *On the Origin of the Species by Natural Selection,* London, Arcturus Publishing Ltd., 2008

80. Google, *Henry Morris quotes,* 2015

81. Google, *Francis Collins quotes,* 2015

82. John Lennox, *Seven Days that Divide the World,* Michigan, Zondervan, 2011

83. Gallup Poll, Google, *Gallup Poll Evolution*, 2015

84. William Paley, *Natural Theology,* UK, Oxford University Press, 2006

85. Donald R Prothero, *Evolution - What the Fossils say and why it matters,* New York, Columbia University Press, 2007

86. Google, *Views on evolution among the public and scientists. NCSE*

87. *God's Undertaker. Has science buried God?* p106

88. Google, *Peter Atkins quotes, 2015*

89. *God's Undertaker. Has science buried God?,* p40

90. *The Grand Design*, p210

91. *God's Undertaker. Has science buried God?,* p44

92. Google, *Karen Armstrong vs Richard Dawkins 2015*

93. Richard Dawkins, *The Blind Watchmaker,* UK, Penguin, 2013

94. *Why There Almost Certainly is a God*, p37

95. *The God Delusion,* p168

96. Collins quote, Google, *God is not threatened /Francis Collins 2015*

97. *The Language of God,* p200

98. John Polkinghorne, Google*, James Gregory lecture, St Andrews, 2008*

99. Richard Swinburne, Google, *Swinburne Is there a God?*

100. Ken Miller, *Exploring the God Question,* DVD series edited by Ian Morris, Programme 2, *Life and Evolution. Google, The God Question*

Chapter 4

101. *Oxford Dictionary,* Google, *Oxford Dictionary*

102. John Lennox, Google, *John Lennox replies to Larry Krauss's claim,* 2015

103. J.D. Watson and F. H. C. Crick, *A Structure for Deoxyribose Nucleic Acid,* Nature, 171, pp737-728, 1953

104. DNA, Google, *Definition of DNA,* 2015

105. Dolly, Google, *The cloning of Dolly the sheep,* 2015

106. HGP, Google, *The Human Genome Project,* 2015

107. HGP, Google, *Initial sequencing and analysis of the human genome,* 2015

108. Nick Bostrom, Google, *The Future of Human Evolution,* 2015

109. Evolution, Google, *Is evolution over? – Instructables,* 2015

110. Future Humans, Google, *Future Humans: Four Ways We May, or May Not Evolve,* 2015

111. Denis Alexander, Google, *Creation, Providence and Evolution/Alexander,* 2015

112. Kelvin, Google, *William Thomson, Lord Kelvin,* 2015

113. Richard Feynman, Google, *Richard Feynman,* 2015

114. Fred Hoyle, Google, *Fred Hoyle,* 2015

115. John Lennox, *God and Stephen Hawking – Whose Design is it Anyway?,* Oxford, Lion Hudson plc, 2011

116. Steven Weinberg, *Dreams of a Final Theory,* London, Vintage, 1994

117. Singularity*, Google, Singularity physics,* 2015

118. Leonard Susskind*,* Google, *The World as hologram,* 2015

119. Brian Greene*,* Google, *Is our Universe the only Universe?* 2015

120. Stephen Hawking, Google, *Hawking makes it clear there is no God,* 2015

121. Erwin Schrodinger, Google, *Schrodinger /quotes* 2015

Chapter 5

122. John Lennox. *God's Undertaker,* p19

123. C S Lewis, *God's Undertaker,* p20

124. Gerald Schroeder, Google, *Schroeder Fine Tuning of the Universe,* 2015

125. Brian Cox, Google, *Cox wonders of the Solar System,* 2015

126. Paul Davies, *The Goldilocks Enigma,* London, Penguin Books, 2007

127. Google, *phase space definition* 2015

128. *God's Undertaker,* p70

129. Roger Penrose, *A Brief History of Time,* (film), Burbank, CA, Paramount Pictures, Inc.

130. Steven Weinberg, *Dreams of a Final Theory. The Search for the Fundamental Laws of Nature,* New York, Vintage, 1994

131. Nancy Frankenberry, *The Faith of Scientists: in their own words,* Princeton University Press, 2008

132. Steven Weinberg, Google, *Weinberg Wikiquotes* 2015

133. Paul Davies, *The Goldilocks Enigma*, p170

134. Roger Penrose, Google, *humanism sir roger penrose*, 2015

135. H. Margenau and R. A. Varghese, ed. 1992, *Cosmos, Bios and Theos. La Salle, IL, Open Court,* p52.

136. *God's Undertaker,* p73

137. Julian Baggini, *Philosophy v Science: which can answer the big questions in life*, The Guardian, Sept. 2012

138. Google, *mission-creep*, 2015

139. *The Language of God,* p3

140. *Exploring the God Question,* DVD series, Google, *the god question*, 2015

141. Loren Haarsma Google, *Haarsma chance, a theistic perspective*, 2015

142. John Polkinghorne, One World: *The interaction of Science and Theology,* Templeton Foundation Press, 2010

143. Albert Einstein, Google, *Einstein Religion and Science, New York Times*

144. Albert Einstein, *The World as I see it,* Citadel Press Inc. US New Addition.

145. Paul Arthur Schilpp, *Albert Einstein: Philosophical Scientist,* The Open Court Publishing Co. La Salle, Illinois, 3rg Ed., 1970

146. Google, *Baruch Spinoza, (Stanford Encyclopedia of Philosophy)* 2001

147. *Quantum Theory. A Very Short Introduction,* p86

148. *God's Undertaker,* pp58-59

149. Google, *20 Amazing Facts about the Human Body, The Guardian*, 2013

150. Brian Greene, YouTube, *The Fabric of the Cosmos – Quantum Leap* p3/4

151. Google, *Hologram, (the Merriam-Webster Dictionary)* 2015

152. YouTube, *Dr Quantum – double slit experiment 2016*

153. Nick Herbert, *Quantum Reality,* Anchor Books, New York, 1985

154. YouTube, *Chaos, It's Only a Theory with Andy Hamilton*, 2016

155. Google, Paul Davies, *Physics and the Mind of God,* Templeton Prize Address, August 1995

156. YouTube, BBC Horizon 2011. *What is Reality?* 2016

157. Google*, Oh this this to solid flesh would melt,* Hamlet soliloquy, 2016

158. Google, The Tempest Act IV scene 1, *We are such stuff,* 2016

159. Google, *How holograms work – explain that stuff,* 2016

160. YouTube, *The Holographic universe,* Part 1 2016

Appendices

161. C. S. Lewis, *Mere Christianity,* W. Collins Sons and Co. Glasgow, 1981

162. *Mere Christianity,* p19

163. *Mere Christianity,* p32

164. *The Language of God*, p2

165. *The Language of God,* p23

166. *Mere Christianity*, p33

167. C. S. Lewis, *The Four Loves,* William Collins Sons and Co, Glasgow, 1979

168. William L. Lorimer, *The New Testament in Scots,* Hazell Watson and Viney Ltd. Aylesbury, Bucks., 1986

169. William Barclay, *The Plain Man Looks at the Beatitudes,* Fontana Books, London, 1963

170. Google, *William Barclay's Daily Study Bible* 2004

171. William Barclay, *The Mind of St Paul,* William Collins Sons and Co., 1983

INDEX

A

Alexander, Denis, 94
Anthropic Principle, 130
Argument from Reason, 15
Atheism, 2, 12, 204
Atkins, Peter, 12, 72

B

Barclay, William, 20, 178, 183, 186, 188, 190
Bible, 3, 20, 63, 65, 84
Black hole, 27, 106, 107, 108
Bohr, Neils, 41, 105, 169
Brian Greene, 44, 109, 153, 165, 166

C

Calvin, Melvin, 120
Chaos Theory, 27, 33, 36, 46 57, 97, 164
Christian, 1, 3, 5, 8, 15, 20, 25, 63, 66, 72, 81, 84, 150, 156, 174, 178, 179, 188, 189
Collider, Hadron, 37
Collins, Francis, 3, 24, 25, 27, 65, 79, 81, 92, 120, 134, 136, 144, 156, 160, 178, 180, 185

Copenhagen Interpretation, 41

D

Dark energy, 27, 30, 33, 53, 56, 106, 117, 127, 129, 143, 161, 164, 171
Dark matter, 27, 30, 53, 54, 56, 58, 106, 127, 128, 164, 171
Darwin, Charles, 12, 62, 90
Davies, Paul, 17, 30, 45, 50, 118, 129, 143, 148, 162, 193, 200, 202
Dawkins, Richard, 2, 3, 5, 10, 11, 13, 14, 18, 21, 23, 27, 31, 59, 70, 72, 78, 102, 134, 149
Dennett, Daniel, 12, 16, 31
Descartes, Rene, 15
DNA, 24, 67, 90, 93, 134, 160, 178
Dualism, 15, 17, 55

E

Einstein, Albert, 6, 25, 35, 39, 97, 139, 140, 161
Electron, 36, 98, 151
Exploring the God Question, 198, 200

F

Fermilab, 37
Feynman, Richard, 97, 152
Fossil, 61, 68, 70, 149, 156

G

God Hypothesis, 11, 17, 23, 24, 173
God of the Gaps, 62, 82, 84
Gribbin, John, 17, 30, 45, 50, 153

H

Hawking, Stephen, 10, 12, 23, 76, 86, 100, 132, 193, 199
Heisenberg, Werner, 39, 41, 155
Hitchens, Christopher, 12
Hoyle, Fred, 45, 127
Human Genome Project, 3, 24, 59, 65, 92, 110, 113, 134, 144, 156, 178

I

Intelligibility, 24, 25, 28, 119, 120, 145, 149, 159, 161, 171

K

Kelvin, Lord, 96, 97, 109

L

Leibniz, Gottfried, 51
Lennox, John, 3, 11, 65, 71, 74, 76, 82, 104, 117, 119, 135, 142, 144, 161, 191, 196, 198
Lewis, C. S., 12, 15, 25, 44, 118, 179, 181, 184

M

Materialism, 5, 7, 14, 17, 26, 32, 45, 54, 59, 88, 100, 150
McGrath, Alister, 3, 8, 11, 12
Mental Matter, 17
Monism, 15, 16, 55
Moral Law, 25, 26, 79, 178, 180, 183

N

Natural Selection, 62, 67, 71, 80, 95, 134, 175
Neuron, 15
Newton, Isaac, 34, 165
Newtonian paradigm, 50
Nonlinear Systems, 48, 49

O

Orzel, Chad, 39

P

Paley, William, 66

Penrose, Roger, 123
Personal Explanation, 14, 23
Physicalism, 44
Planck, Max, 12, 38, 148, 155
Polkinghorne, John, 11, 25, 39, 43, 86, 131, 136, 147, 156, 174, 176
Prothero, Donald, 68, 70

Q

Quantum physics, 6, 9, 12, 34, 38, 40, 45, 47, 97, 104, 115, 152, 155, 164, 169, 171

S

Schroedinger, Erwin, 115
Seth Lloyd, 4, 34
Sheldrake, Richard, 17, 56, 98, 101, 107, 167
Singularity, 107, 167
Soliton, 48
Spinoza, Baruch, 51, 140
Standard Model, 38, 54
Supernatural, 3, 9, 14, 19, 28, 78, 111, 120, 149, 159, 162, 170
Susskind, Leonard, 107, 166, 168

Swinburne, Richard, 93

T

Theistic Evolution, 6, 62, 64, 78, 80, 83, 85, 155, 170
Theory of everything, 100, 108

U

Ultimate reality, 9, 11, 28, 32, 42, 54, 57, 106

V

Virtual particle, 42

W

Ward, Keith, 3, 10, 12, 16, 23, 37, 78
Watson and Crick, 96
Weinberg, Steven, 103, 129, 130

Y

YEC, 64, 66, 156

34069905R00120